AF078949

SPRINGBOARD
KS3 SCIENCE

PRACTICE BOOK 2

Claudia Allan
Jovita Castelino
Thomas Millichamp
Adam Robbins
Bill Wilkinson

SERIES EDITOR
Adam Boxer

Although every effort has been made to ensure that website addresses are correct at time of going to press, Hodder Education cannot be held responsible for the content of any website mentioned in this book. It is sometimes possible to find a relocated web page by typing in the address of the home page for a website in the URL window of your browser.

Hachette UK's policy is to use papers that are natural, renewable and recyclable products and made from wood grown in well-managed forests and other controlled sources. The logging and manufacturing processes are expected to conform to the environmental regulations of the country of origin.

To order, please visit www.hoddereducation.com or contact Customer Service at education@hachette.co.uk / +44 (0)1235 827827.

ISBN: 978 1 3983 8575 7

© Claudia Allan, Adam Boxer, Jovita Castelino, Thomas Millichamp, Adam Robbins, Bill Wilkinson 2024

First published in 2024 by
Hodder Education
An Hachette UK Company
Carmelite House
50 Victoria Embankment
London EC4Y 0DZ

www.hoddereducation.com

The authorised representative in the EEA is Hachette Ireland,
8 Castlecourt Centre, Dublin 15, D15 XTP3, Ireland (email: info@hbgi.ie)

Impression number   10 9 8 7 6 5 4

Year   2028 2027 2026 2025

All rights reserved. Apart from any use permitted under UK copyright law, no part of this publication may be reproduced or transmitted in any form or by any means, electronic or mechanical, including photocopying and recording, or held within any information storage and retrieval system, without permission in writing from the publisher or under licence from the Copyright Licensing Agency Limited. Further details of such licences (for reprographic reproduction) may be obtained from the Copyright Licensing Agency Limited, www.cla.co.uk

Cover illustration © Sara Lynn Cramb

Typeset in India

Printed in Great Britain by Bell and Bain Ltd, Glasgow

A catalogue record for this title is available from the British Library.

CONTENTS

Get the most from this book　　5

BIOLOGY

B3　Nutrition and digestion　　6
　B3.1　Healthy diet, energy requirements and dietary imbalance　　6
　B3.2　Digestive organs　　12
　B3.3　Gut bacteria　　16

B4　Gas exchange systems　　20
　B4.1　Ventilation　　20
　B4.2　Gas exchange　　25
　B4.3　Exercise, asthma and smoking　　27

B5　Reproduction　　34
　B5.1　Sexual reproduction and the reproductive organs　　34
　B5.2　Fertilisation　　37
　B5.3　Fetal development　　39
　B5.4　The menstrual cycle　　45
　B5.5　Plant reproduction　　48

CHEMISTRY

C4　Chemical reactions　　53
　C4.1　Reactions, conventions and signs a reaction has occurred　　53
　C4.2　Combustion, thermal decomposition, oxidation and displacement　　57
　C4.3　Conservation of mass　　61
　C4.4　Acids, alkalis and pH　　65
　C4.5　Reactions of acids with metals and alkalis　　68

C5　Energy changes　　73
　C5.1　Changes of state　　73
　C5.2　Endothermic and exothermic reactions　　76

PHYSICS

P4 Pressure in fluids — 82
- **P4.1** Pressure in liquids — 82
- **P4.2** Atmospheric pressure — 84
- **P4.3** Pressure calculations — 86

P5 Sound — 90
- **P5.1** Types of wave — 90
- **P5.2** Sound waves — 95
- **P5.3** Microphones and ultrasound — 103

P6 Light — 106
- **P6.1** Light and ray models — 106
- **P6.2** Interactions of light waves with materials — 112
- **P6.3** Mirrors, pinhole cameras and the eye — 117
- **P6.4** Detecting light and colour — 122

Get the most from this book

Welcome to Springboard: KS3 Science!

Springboard: KS3 Science has been developed by our expert author team to help you build, develop and sustain your knowledge and understanding of science through your first three years at secondary school and to inspire you to be a lifelong scientist.

Your Practice Books provide you with lots of questions to help you learn, remember and apply everything your teacher has taught you.

Each topic includes questions related to earlier topics, too. These questions help you to recall what you have learned and to make connections between different ideas, as you build your scientific understanding of the world around you.

Lots of questions on each topic are available to help you really fix what you have learned into your memory. This will help you to remember and apply it for longer.

Tips offer hints about key information you can find in earlier topics. You might need to look back to earlier topics in your Knowledge Book if you do not remember the information you have learned.

Your teacher has access to answers for all the Practice Book questions in Boost and will review your answers with you.

Tips remind you to look at Worked examples in your Knowledge Book. These Worked examples give you step-by-step solutions to questions that use key skills covered in a topic. They show you how to approach similar questions in your Practice Books.

Get the most from this book 5

B3 Nutrition and digestion

B3.1 Healthy diet, energy requirements and dietary imbalance

What are the components of a healthy diet?

1. Why must we include carbohydrates in our diet?
2. True or false: fruits contain all the key nutrients required to be healthy. Give a reason for your answer.
3. What nutrient groups do cheese, pasta and eggs provide?
4. Why are fats needed in the diet?
5. What foods could you eat to give yourself more energy?
6. Name some foods a person should avoid if they are trying to reduce their fat intake.
7. What nutrients does fish provide?
8. Why are proteins needed in the body?
9. Copy and complete the following sentences.
   a Fats are needed in the diet because …
   b Fats are needed in the diet, but …
   c Fats are needed in the diet, so …
10. The card below shows the amount of fat and fibre in some types of food and drink from a café.

	fat in g	fibre in g
type of drink		
chocolate milkshake	7	0
lemonade	0	0
orange juice	0	0
type of burger		
single burger	15	0.7
double burger	36	1.1
cheeseburger	22	0.9
type of potato		
French fries	16	4
baked potato	0	8

   a From the card, choose a meal consisting of a burger, a drink and some potato, to give:
   i the least fat
   ii the most fibre.

b A person orders a double burger, French fries and a chocolate milkshake. Calculate the fat content of this meal. Give the unit.

11 Which nutrient do we need to grow and repair tissues?

12 A student says, 'A healthy diet must only include fruits and vegetables.' Explain why they are wrong.

13 A person has a headache and is told to drink more water. Why do our bodies need water?

14 In what part of the cell do chemical reactions take place?

15 A student wants to investigate how much energy is given out when burning different amounts of bread.

    a Write a line of inquiry for this experiment.

    b Which is the independent variable?

    c Which is the dependent variable?

    d What are some control variables they need to consider?

16 A healthy diet is needed to build muscle. This question is about muscles.

    a Name parts A, B and C of the muscle cell in the diagram below.

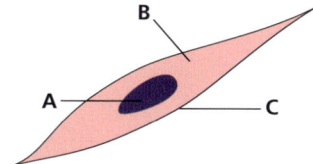

    b A student says the diagram shows a plant cell. Explain why they are wrong.

    c What are antagonistic muscles?

    d This cell is 0.007 cm in diameter. How big will it appear to be if it is magnified ×600? Use the EVERY method to show your working.

    e Convert your answer to millimetres.

    f What is the function of muscles?

> **Tip**
> For help, see the Worked example in Topic B1.4 of your Knowledge Book.

17 A plant cell is different to a muscle cell as it has a cell wall. What is the function of the cell wall in plant cells?

18 Plants also require certain nutrients to survive. They can make their own food through the process of photosynthesis. What is the name of the sub-cellular structure responsible for photosynthesis in plant cells?

19 What is the function of the mitochondria in animal cells?

20 What is the shape of red blood cells?

21 What is the function of red blood cells?

22 Which sub-cellular structure is missing in red blood cells?

B3 Nutrition and digestion

How do we draw bar charts?

23 Why do organisms need energy?

24 A 9-month-old baby needs approximately 700 calories a day. A 2-year-old needs approximately 1200 calories a day. Give two reasons why the 2-year-old needs more calories.

Tip

These questions also cover 'Why do some people need more food than others?'

25 A scientist is interested in how much energy is consumed by a group of people. The bar chart below shows their results.

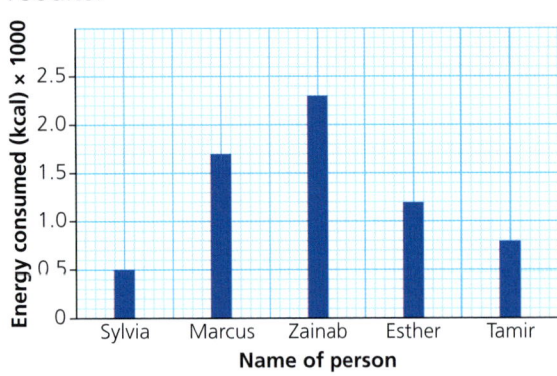

 a Which of these people is most likely the youngest?
 b Explain why Zainab is the most likely to be an athlete.
 c Which nutrient groups provide energy?
 d Why do we need energy?

26 A student wants to find out if the amount of protein a person eats affects their muscle strength.

 a Write a line of inquiry for this investigation.
 b What is the independent variable?
 c What is the dependent variable?
 d What are two control variables?
 e Why do we need proteins in our diet?
 f What are two foods that provide protein in our diet?
 g The graph below shows the student's results. What is the relationship between mass of protein and muscle strength?

Tip

For help, see the 'Working scientifically' box in the 'Lines of inquiry and variables' section of your Knowledge Book.

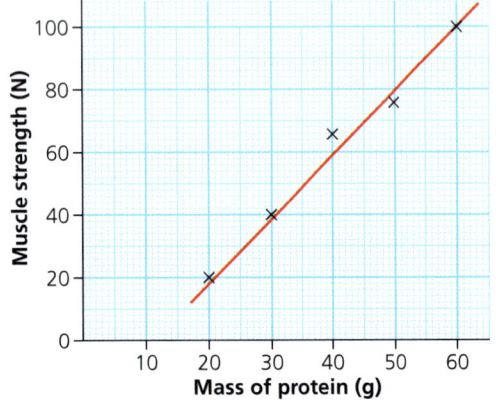

27 The table below shows results from an experiment conducted by a scientist.

Age group (years)	Energy consumed per day (kJ) × 1000
0–5	3
6–10	5
11–15	8
16–20	11
21–25	14

a Write a line of inquiry for these results.

b What type of graph should the scientist draw to represent these results?

c Explain your answer to part b.

28 The bar chart below shows the amount of energy required each day by different people.

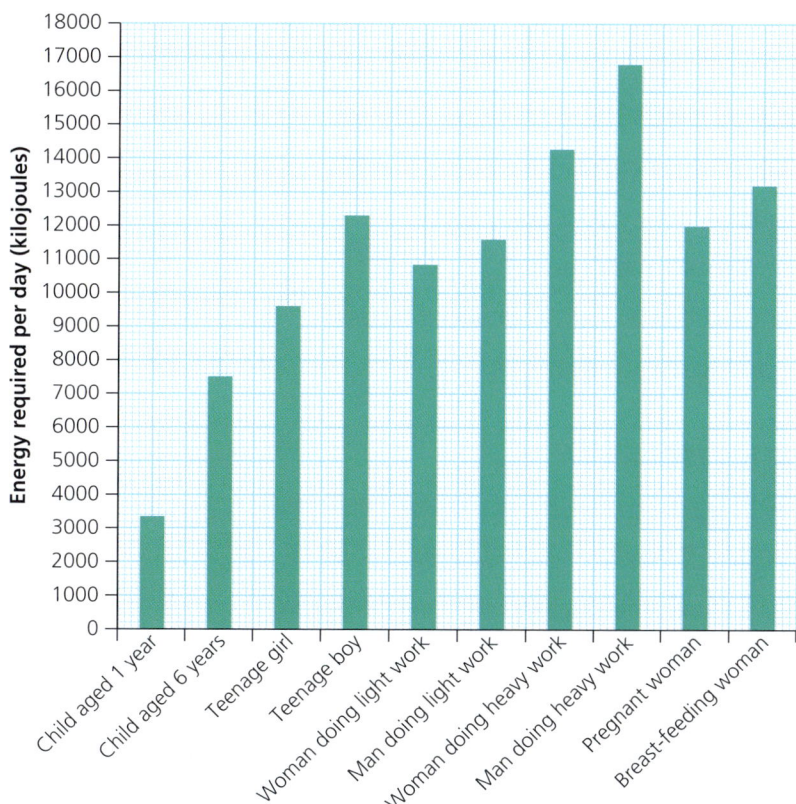

a Which category of people requires the most energy per day?

b How does energy requirement per day change with age?

c How much energy does a woman doing light work require in a week? (A week has 7 days.)

d Why does a man doing heavy work require more energy than a man doing light work?

B3 Nutrition and digestion

e A man doing office work changes career to become a gardener. How much more energy will they require per day? Use the average values from the bar chart.

29 What type of chart is used to compare energy requirements of different people?

30 A student says plants do not need any energy because they do not move. Explain why they are wrong.

31 What is the function of ribosomes in cells?

32 What type of muscle cells are responsible for involuntary movements such as heartbeats?

33 What type of muscle cells are responsible for voluntary movements such as walking and talking?

What are diet-related diseases?

34 Someone with obesity is most likely consuming foods from which nutrient group?

35 How could someone avoid getting anaemia?

36 What can cause heart disease?

37 Why is fibre needed in a healthy diet?

38 A person feels tired all the time and has shortness of breath.
   a What deficiency disease have they most likely got?
   b What change to their diet should they make?
   c Suggest foods that they should eat to help them feel better.

39 What happens when too many fatty substances get deposited in arteries?

40 What can result from eating too many fats?

41 What is a deficiency disease?

42 What are proteins used for in the body?

43 Babies and infants are given lots of milk as part of their diet. Why do you think this is?

44 Name two conditions linked with unbalanced energy intake.

45 Each statement below is incorrect. Write a correct version of each statement.
   – People who need to lose weight need to take in less protein than they use.
   – The nutrient groups are carbohydrates, fats, proteins, dairy, vitamins, minerals and water.
   – All people of the same age should consume the same amount of energy.

46 This question is about red blood cells.
   a What is the function of red blood cells?
   b How is the red blood cell adapted to its function?

c In which deficiency disease are red blood cells affected?

d What are the symptoms of this disease?

e A student knows that a red blood cell is 0.065 mm long. How long will it appear to be when magnified ×350?

47 This question is about fats.

a Why do we need fats in our diet?

b Fats are stored in fat cells. Name the five sub-cellular structures found in animal cells.

c Which sub-cellular structure releases energy?

48 Look at the bar chart below.

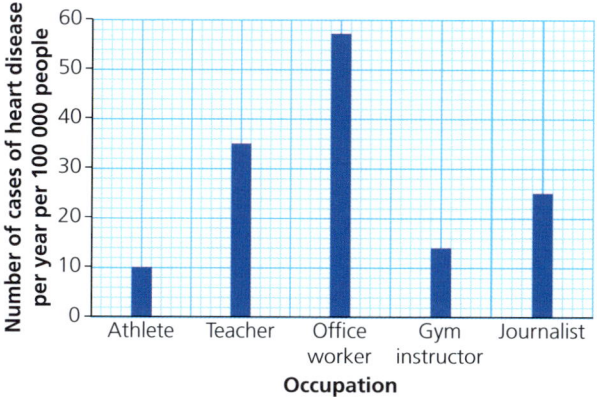

a Write a line of inquiry for these results.

b Which occupation has the most cases of heart disease per year?

c Which occupation has the least cases of heart disease per year?

49 This question is about athletes.

a How does the energy requirement of an athlete differ from that of the person doing the commentary during the race?

b What is the change in energy stores that takes place for an athlete during a race?

c In what way is the energy transferred between the two stores as the athlete races?

d Athletes need to grow and repair muscles. Which nutrient group do they need more of for this?

e Athletes are often more muscular than non-athletes. What is the function of muscles?

f What is a muscle tissue?

g When an athlete is running, they use their hamstrings and quadriceps muscles in their legs. What name is given to this muscle pair in terms of how they work together?

h When an athlete bends their knee to run, the hamstrings contract. What happens to the quadriceps?

B3 Nutrition and digestion

50 Which nutrient group is important for maintaining healthy digestion and preventing constipation?

51 Which nutrient group is an important source of insulation and cushioning for organs?

52 What is the name of the tissue that connects bones to other bones?

53 What is the name of the tissue that connects muscles to bones?

54 What is the name of the mineral that is important for healthy blood and oxygen transport?

B3.2 Digestive organs

Why do we digest food?

55 What is digestion?

56 Why is digestion important to us?

57 Why do we need vitamins in our diet?

58 What are enzymes?

59 Why does food need to be digested?

60 This question is about proteins.

    a Why do we need proteins in our diet?

    b Which sub-cellular structure allows proteins to be made?

    c Lots of proteins are made in the liver. A liver cell measuring 0.0032 mm long is magnified ×1200. How long will it appear to be after magnification?

    d A scientist has investigated how the mass of proteins eaten affects muscle mass. The graph below shows their results. What does the graph show?

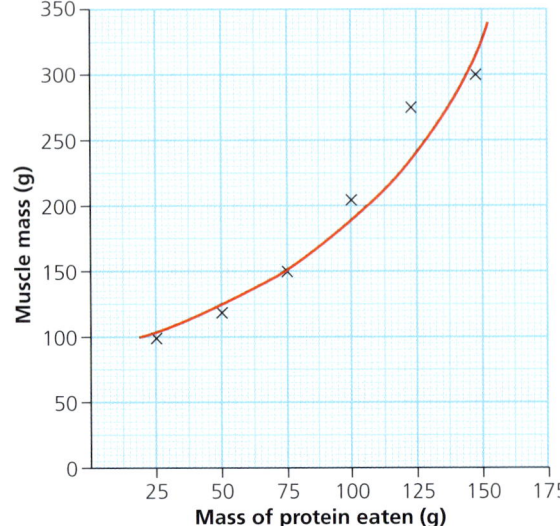

61 Give two symptoms of being underweight.

62 In which parts of the digestive system are enzymes produced?

63 A student has written, 'Anaemia is caused by a lack of iron. Iron is needed to carry oxygen around the body so people with anaemia feel tired all the time.' Rewrite their answer to include the correct detail.

64 Why do we need water in our diet?

65 What is obesity?

66 Which nutrient group is the body's main source of energy?

67 Which nutrient group is lacking if a person cannot repair their tissues?

68 Hormones are made of proteins. In which part of the cell are proteins made?

69 Name three food items a person should eat to get lots of protein.

70 Which type of muscle surrounds organs in the body?

71 A person takes up dancing. Which nutrient group do they need more of so they can move around more?

72 Why does the person who has started dancing need more energy than before?

What path does food take when we eat?

73 Name parts A to I in the diagram of the digestive system below.

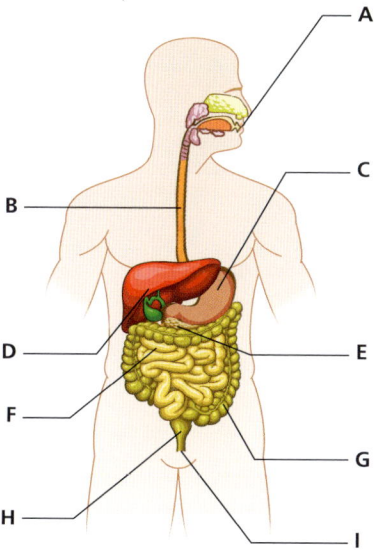

74 What is the difference in function of the small and large intestines?

B3 Nutrition and digestion

75. What is the role of saliva in the mouth?
76. Why do we need proteins in our diet?
77. Where is bile produced?
78. What do enzymes do?
79. Why is the stomach an organ?
80. Why does food need to be digested?
81. A student says, 'The liver is a part of the digestive system, even though food does not pass directly through it.' Explain why they are correct.
82. A student says, 'Most of the process of digestion takes place in the stomach.' Explain why they are wrong. Refer to specific parts of the digestive system in your answer.
83. What are the seven nutrient groups?
84. A muscle cell from the stomach is 0.4 mm wide. How big will it appear to be when magnified ×1500?
85. Compare and contrast the functions of the mouth and the stomach.
86. Some people struggle to chew food. Why do these people often have to have their food blended and given to them as a liquid?
87. A gastric band is a band that goes around the stomach. It means that a person's stomach has a smaller volume and they feel full more quickly. Explain why this could cause a person to lose weight.

How is the small intestine adapted to its function?

88. Explain how the small intestine is adapted for efficient absorption of digested food molecules.
89. A person has a condition in which their small intestine is covered in fewer villi compared to most people. How will this affect the function of the small intestine?
90. Some enzymes only work well in acid. Where in the body do you think these enzymes are found?
91. A student eats a bread roll. Explain what happens to this bread roll as they chew and digest it.
92. What is diffusion?
93. Which part of the digestive system helps to kill bacteria found in food?
94. A student is suffering from diarrhoea or watery faeces. Which part of the digestive system is not functioning correctly? Explain your answer.

95 What are the two roles of the small intestine?

96 A scientist wants to investigate if the type of medicine taken affects the mass of sugars absorbed by the small intestine.

    a Write a line of inquiry for this investigation.

    b What is the independent variable in this investigation?

    c What is the dependent variable in this investigation?

    d The scientist makes sure all the people they are studying are consuming (eating) the same mass of sugar. What type of variable is this?

97 Sugars are a type of carbohydrate. Why do we need carbohydrates in our diet?

98 Copy and complete the following sentences.

    a The small intestine has a large surface area because …

    b The small intestine has a large surface area, so …

99 The diagram below shows a cell.

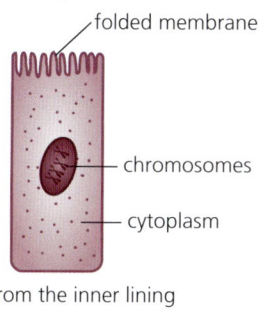

cell from the inner lining of the intestine

    a What is the function of the cytoplasm?

    b The width of one microvillus is 0.0002 mm. If it is magnified ×1400, how wide will it appear to be?

    c What is the total width of the microvilli of this cell?

    d What is the process by which food molecules pass through the membranes and into the bloodstream?

100 Where is bile produced?

101 Correct this statement: All people require the same amount of energy to be healthy.

102 Obesity increases the risk of which three diseases?

103 Why do we need vitamin C in our diet?

104 This question is about blood.

    a Where are blood cells made in the human body?

    b What is the function of the red blood cells?

    c Why do red blood cells have a biconcave shape?

    d What mineral is needed for red blood cells to function well?

    e Name the disease caused by a lack of this mineral.

105 What is the role of the digestive system?

106 What are enzymes made up of?

107 What is the name of the process by which larger carbohydrates are broken down into smaller ones?

108 What is the name of the muscular tube that connects the mouth to the stomach?

109 What is the name of the opening at the end of the digestive system?

110 A student says a person who is obese has a lot of fat in their stomach. Explain why they are wrong.

111 Where are enzymes made within a cell? Explain your answer.

112 What would happen to the function of the large intestine if it had several folds within its structure? Explain your answer.

113 What would happen to the function of the small intestine if it had a smooth, flat membrane? Explain your answer.

114 What is the function of the cell membrane?

115 A shampoo bottle says it is protein rich. Explain why applying this shampoo on hair does not provide the proteins needed for growth and repair of intestine cells.

B3.3 Gut bacteria

116 A student says that all bacteria are harmful. Explain why they are wrong.

117 Copy and complete the following sentences.

   a Gut bacteria are helpful, because …

   b Gut bacteria are helpful, but …

   c Gut bacteria are helpful, so …

118 Probiotics are foods which contain bacteria that can add to your gut bacteria. Some yoghurts are probiotics. Explain why eating yoghurt might have a positive effect on a person's health.

119 Why is the genetic material of bacterial cells found in the cytoplasm?

120 Many yoghurts contain lots of sugar and fat. How could eating these types of yoghurt affect a person?

121 What are the main adaptations of the small intestine to help it absorb food molecules?

122 A student is found to have too many harmful bacteria in their gut. Suggest what foods they should add to their diet to help reduce harmful bacteria in their gut.

123 In a scientific study, some people changed their diet to include more burgers, pizzas and chips. The scientists found that this had a negative effect on gut bacteria. What conclusion can be made from this about digestion and diet?

124 Give two ways the bacteria in our gut help us.

125 What structure of a bacterial cell protects it?

126 Why do we need fibre in our diet?

127 Where are enzymes released from in the digestive system?

128 Why do we need minerals in our diet?

129 A student follows a vegan diet. This means they do not eat meat and dairy products. Recently, they have been feeling very tried, short of breath and lacking in energy.

    a What deficiency disease could they have?

    b What causes this disease?

    c What change should they make to their diet to improve their symptoms?

130 What is the cause of heart disease?

131 Why do athletes need more energy than people who are not as active?

132 What are the functions of the stomach in digestion?

133 Which part of the digestive system is involved in absorbing nutrients?

134 Compare and contrast the structure of a bacterial cell and a plant cell.

135 A student is underweight. What symptoms might they have?

136 A person has a high-fat diet. What are two problems they may have with their health?

137 Why does the small intestine have many finger-like projections instead of a smooth surface?

138 Rishab has just eaten a meal and is doing a handstand. His friend Karan says that his food will come back down into his mouth instead of going to his stomach. Explain why Karan is wrong.

139 Why do we digest food?

140 What can a build-up of fatty deposits in blood vessels lead to?

141 State all the parts of the digestive system in the order in which food travels through them.

142 An adult gets 10 000 kJ of energy every day from their food. They exercise and use 9200 kJ of energy. What could happen to that person? Explain your answer.

B3 Nutrition and digestion

143 Read the extract below about research into heart disease.

More heart disease in older women: Heart disease among British women in the 60–79 age group is more common than previous research suggested. A recent study of 4286 British women in that age group indicated that one in five showed signs of heart disease.

  a Why can the results of this research not be used to draw any conclusions about heart disease among women across the world?

  b How can people reduce their risk of heart disease?

144 The table below shows some information about a packet of biscuits.

	Average values		UK guideline daily amounts	
	Per 100 g	Per biscuit	Adults	Children (5–10 years)
Energy	1974 kJ	446 kJ	8500 kJ	7500 kJ
Proteins	7.1 g	1.1 g	45 g	24 g
Carbohydrates	62.8 g	9.3 g	230 g	220 g
Fats	21.3 g	3.2 g	70 g	70 g
Sodium	3.6 g	0.5 g	2.4 g	1.4 g

  a One day, a 10-year-old child ate a whole packet of biscuits. The biscuits in the pack had a mass of 400 g. How many grams of carbohydrates did the child eat?

  b The amount of carbohydrates you calculated in part a was more than the UK guideline daily amount for the child. How much more?

  c How many biscuits would an adult need to eat to get all the protein they need?

  d How much energy would that number of biscuits give them?

  e Is eating biscuits a good way to get protein? Explain your answer.

  f An adult eats three biscuits. What percentage of their daily amount of fat is that?

145 What will happen to a person who must have 2000 kcal of energy every day but only consumes foods containing 1800 kcal each day?

146 Which organ absorbs water from digested food into the bloodstream?

147 Why do humans have bones?

148 What is the difference between ligaments and tendons?

149 Which function of the skeleton allows oxygen to be transported around the body?

150 Calcium is important for building bones. Which nutrient group does calcium belong to?

151 A 0.005 mm wide bone cell is observed under a microscope with ×500 magnification. How wide does it appear under the microscope?

152 A student says all bacteria make us ill. Explain why they are wrong.

153 The table below shows sizes of different cells and sub-cellular structures. Which sub-cellular structures would fit within a bacterial cell?

Cell/sub-cellular structure	Size (micrometres)
Bacterial cell	0.5
Animal cell	10
Plant cell	10
Mitochondrion	2
Ribosome	0.005
Nucleus	5
Permanent vacuole	0.1
Chloroplast	2

B4 Gas exchange systems

B4.1 Ventilation

What is the respiratory system?

1. What is ventilation?
2. Why do we need to breathe?
3. Copy and complete the table using the diagram below.

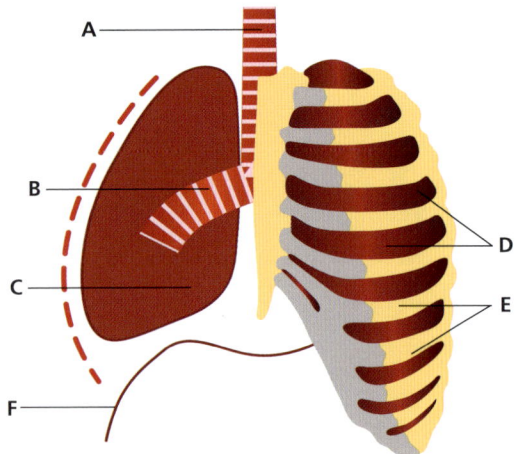

Label	Name
A	
B	
C	
D	
E	
F	

4. What tissue is the diaphragm made of?
5. What is the function of the cartilage rings on the trachea?
6. A student says when we breathe, we only take in oxygen. Explain why they are incorrect.
7. What is the role of the rib cage?
8. Name the two main muscles that are part of the respiratory system.

9 What is the name given to the process by which gases enter and leave the lungs?

10 A person has a collapsed trachea, which means their trachea is bent inwards under pressure. What part of the trachea has been broken?

11 Name two types of cell that line the inside of the trachea.

12 Cystic fibrosis is a disease where a person makes too much mucus in their respiratory system. What part of the respiratory system is involved in making mucus?

13 How do the cells lining the trachea work together to get rid of dust and bacteria?

14 This question is about cilia.
   a What is the function of cilia in the trachea?
   b The length of 50 cilia is 0.6 mm. How long will they appear to be if they are magnified ×150?
   c How long will one of the cilia appear at this magnification?
   d Cilia are found on ciliated epithelial cells. What is the function of the nucleus within the cells?
   e A student says that ciliated epithelial cells and bacterial cells are identical because they both have tail-like structures. Explain why they are wrong.

> **Tip**
> The question is asking how long all 50 cilia will appear.

> **Tip**
> If 50 cilia are as long as you found in part b, how long will one of the cilia be?

15 A student wants to investigate which type of activity affects breathing rate the most. They look at three types of activity: walking, running and hopping.
   a Write a line of inquiry for this investigation.
   b What is the independent variable?
   c What is the dependent variable?
   d Suggest two control variables.
   e A person takes 17 breaths in 20 seconds. If they keep breathing at this rate, how many breaths would they take in 1 minute?

> **Tip**
> For help, see the 'Working scientifically' box in the 'Lines of inquiry and variables' section of your Knowledge Book.

> **Tip**
> Remember, there are 60 seconds in 1 minute.

16 Fix this incorrect statement: 'The mitochondria are a type of muscle cell.'

17 The ribs are part of the skeleton. There are several joints within the skeleton. What is the definition of a joint?

18 True or false: the cell membrane is the outermost part of the cell.

19 What is the purpose of the ciliated epithelial cells?

20 In which part of the human body are goblet cells found?

B4

How do the lungs work?

21 The diagram below shows a bell jar model.

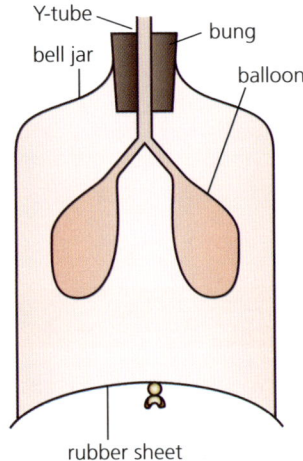

a Which part of the bell jar model represents the:

   i trachea   **iii** diaphragm

   ii lungs   **iv** rib cage?

b Name two structures that are not represented by the bell jar model.

22 Copy and complete the table below.

	Inhaling	Exhaling
Intercostal muscles		
Ribs		
Diaphragm		
Lung volume		
Lung pressure		

23 The diagram below shows cells found in the trachea.

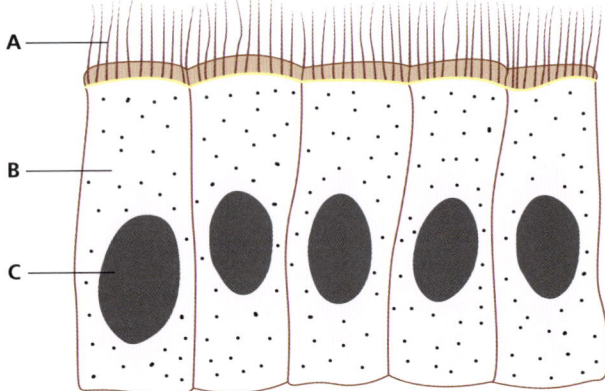

a What is the name given to this type of cell?

b Name the sub-cellular structures A–C.

> **Tip**
>
> Remember the basic structure of a cell and use that knowledge to identify these structures.

24 What is ventilation in the respiratory system?

25 What happens to the diaphragm when we inhale air?

26 What happens to the intercostal muscles when we exhale air?

27 What is the path taken by air when we breathe in?

28 This question is about muscles.
   a Name the three types of muscle found in the human body.
   b What is the function of muscle cells?
   c What is a muscular tissue?
   d What sub-cellular structure is present in large numbers in muscle cells?
   e Muscles need to repair themselves often. What nutrient group is required to help with this?

29 This question is about alveoli.
   a A microscope is used to observe alveoli. The eyepiece magnification is ×20, and a ×50 objective lens is used. What is the total magnification?
   b The width of one alveolus is 0.0006 mm. What is the width of the image when magnified using this microscope?
   c What part of the respiratory system leads directly to alveoli?

30 The ribs are made of bone. Give the four functions of the skeleton.

31 This question is about cartilage.
   a What is the function of cartilage in joints?
   b Where is cartilage found in the respiratory system?
   c What is the function of cartilage in the thorax?
   d The width of the trachea without cartilage is 9 mm and with cartilage is 20 mm. How many times wider is the trachea with cartilage than the trachea without cartilage?

32 This question is about bacteria.
   a Give two differences in the structures of bacterial and animal cells.
   b How does the respiratory system protect us from harmful bacteria?
   c In what two ways does our digestive system protect us from harmful bacteria?

33 Copy and complete the following sentences.

  a The trachea has ciliated epithelial cells because …

  b The trachea has ciliated epithelial cells, but …

  c The trachea has ciliated epithelial cells, so …

34 Explain what happens within the respiratory system when we inhale.

35 Why are the lungs considered to be an organ?

36 A scientist wants to know if the number of alveoli a person has in one lung is related to their lung volume.

  a Write a line of inquiry for this investigation.

  b What is the independent variable?

  c What is the dependent variable?

  d Name two control variables.

  e The scientist's results are represented in the graph below. What is the relationship between the number of alveoli in one lung and the lung volume?

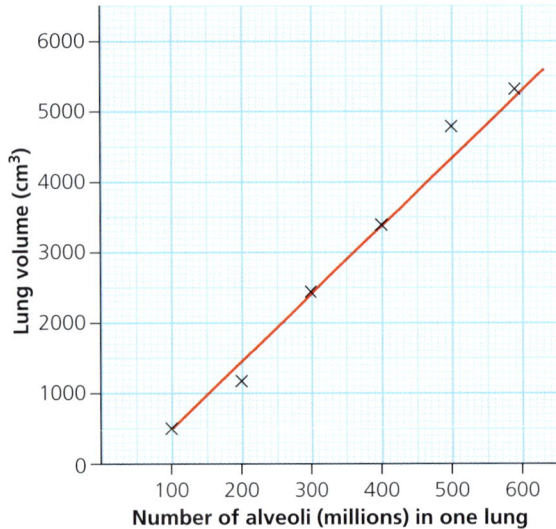

37 Fix this incorrect statement: 'Inhaling and ventilation are the same thing.'

38 Scientists can study cells within the lungs using a microscope, which magnifies an image. What is magnification?

39 What is the diaphragm made of?

40 Which organ system is the trachea a part of?

41 A person has a condition which means they cannot repair bone cells. What part of their body will be affected by this condition?

B4.2 Gas exchange

How does gas exchange happen?

42 Copy and complete the table below to summarise what happens during gas exchange in the lungs.

	Oxygen	Carbon dioxide
Where is the concentration the highest?	Alveoli	Bloodstream
Where is the concentration the lowest?		
Which way will the gas move?		
By which process will the gas move?		

43 Which gas moves from the alveoli into the blood?

44 Which gas moves out of the blood into the alveoli?

45 Describe what happens to oxygen and carbon dioxide inside the alveoli.

46 The oxygen-rich blood leaves the lungs and is returned to the heart to be pumped around the body. As it travels around the body, the blood passes cells that are low in oxygen. What do you think happens?

47 This question is about the cells that make up alveoli.

  a An alveolar cell is magnified ×1500. If it is 0.000 75 mm wide, how big does it appear when magnified?

  b What is the function of the cell membrane?

  c What is diffusion?

  d A student says that all the gases we inhale diffuse into our capillaries. Explain why they are wrong.

48 What is gas exchange in the lungs?

49 Why does the trachea have ciliated epithelial cells?

50 What happens to the oxygen that crosses into the bloodstream?

51 Why does air rush in when we inhale? Include a description of changes at the ribs, intercostal muscles, diaphragm and lungs in your answer.

52 A scientist is investigating if the temperature within the alveoli affects the rate of diffusion of oxygen into alveoli.

  a Write a line of inquiry for this investigation.

  b What is the independent variable?

  c What is the dependent variable?

  d What are two control variables?

> **Tip**
>
> These questions also cover 'What is the composition of the air that we inhale and exhale?'

> **Tip**
>
> Even if the numbers are very small or very large, follow the EVERY method to calculate the correct answer.

B4 Gas exchange systems

e Predict what effect the temperature within the alveoli will have on the rate of diffusion of oxygen. Explain your answer.

53 True or false: the higher the magnification, the wider the field of view.

54 In which structure of the lungs does gas exchange take place in humans?

55 What is the process of exchanging oxygen and carbon dioxide between the lungs and blood called?

56 Fix this incorrect statement: The main function of goblet cells is to provide structure to the lungs.

57 What are the hair-like structures found in the trachea called?

58 A student says we die if we do not take all the oxygen from the air we inhale into our blood. Explain why they are wrong.

What are the adaptations of the gas exchange system?

59 Which gas diffuses from the blood into the alveoli?

60 What is the name of the sheet of muscle separating the thorax and the abdomen?

61 Give two ways in which the alveoli are adapted for their function.

62 How do the following features of the respiratory system make them good at gas exchange?

  a Rings of cartilage in the trachea

  b Thin membranes of the alveoli

  c Highly folded alveoli membranes

  d A good blood supply around the alveoli

63 What would happen to the mucus in the trachea if the ciliated epithelial cells did not have cilia?

64 Name three factors that affect the rate of gas exchange in the lungs.

65 When do the lungs have higher pressure: during inhaling or exhaling?

66 Give two differences between the composition of inhaled and exhaled air.

67 Give one similarity between the composition of inhaled and exhaled air.

68 Why does exhaled air contain less oxygen than inhaled air?

69 Why does oxygen diffuse into the blood during gas exchange?

70 What happens to the pressure in the lungs as the chest volume decreases?

71 What shape are the cells of the alveoli?

72 What is increased by the shape of the alveoli?

73 Why is it important that the membranes of the alveoli and capillaries are only one-cell thick?

74 What are the main structures found in the gas exchange system?

75 How would an alveolus with very few folds compare in terms of gas exchange to one with a highly folded membrane?

76 A student says the alveoli are where gases move from the lungs to the blood. Improve their answer to include detail about gas exchange.

77 Give two similarities between the small intestine and the alveoli.

78 What would happen to the function of alveoli if they had smooth, flat membranes? Explain your answer.

79 Through what structure does air enter the alveoli?

80 Intercostal muscles are found between ribs. What sub-cellular structure is found in large quantities in the cells making up these muscles? Explain your answer.

81 What does the word *epithelial* mean?

82 Do we exhale more or less carbon dioxide than when we inhale?

83 Why is it important for alveoli to be moist?

B4.3 Exercise, asthma and smoking

How does exercise affect breathing?

84 What percentage of oxygen is present in inhaled air?

85 Why do we breathe in more during exercise?

86 This question is about exercise.

   a What effect does exercise have on breathing rate?

   b What gases are exchanged in the alveoli?

   c What do our muscles do when we exercise?

   d What adaptations do muscle cells have to help them with their function?

87 What happens to the respiratory system when we exhale?

88 A student says all our muscles contract when we exercise. Explain why they are wrong.

89 What is ventilation?

90 Fraser and Ryan wanted to find out which of them is fitter. They both did some exercise which made them breathe faster. They then stopped exercising and timed how long it took for their breathing rates to return to normal again. The graph below shows the results.

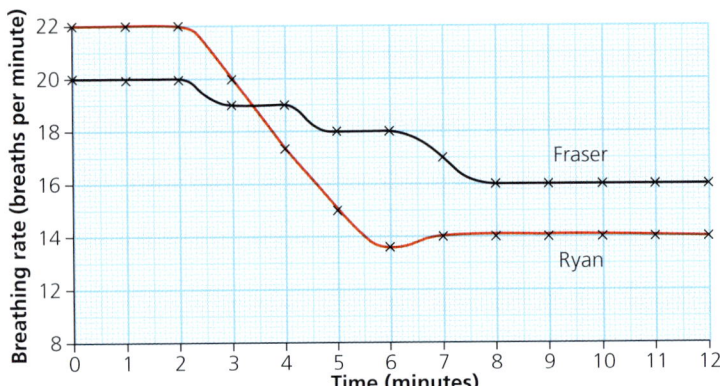

   a Identify an error in the scale of the graph.

   b After how many minutes did Fraser and Ryan stop exercising?

   c Ryan said that the faster a person's breathing rate returned to a steady resting rate, the fitter the person was.

   From the information in the graph, state who was the fitter person and explain why.

91 In terms of gases, what changes occur in cells during exercise?

92 A person has arthritis. This means they have painful joints.

   a What is a joint?

   b The person struggles to exercise. Suggest how this affects the muscles in their thorax compared to when they were able to exercise before developing arthritis.

93 What are four adaptations of the gas exchange system?

94 The alveoli have walls that are one-cell thick.

   a What is the function of the cytoplasm?

   b Which sub-cellular structure allows gases to diffuse into and out of the cell?

   c An alveolar cell is 0.004 mm long. How long will it appear to be if it is magnified ×100?

   d What is diffusion?

   e Why do alveoli have walls that are one-cell thick?

95 This question is about muscles.

   a Why do muscle cells need proteins?

   b Which sub-cellular structure provides the instructions to make proteins in muscle cells?

    c Which sub-cellular structure releases energy in muscle cells?

    d If the thorax did not have intercostal muscles, what effect would this have on ventilation?

96 A person has fewer alveoli than normal. How will this affect:

    a the surface area of the gas exchange system

    b gas exchange compared to others

    c their breathing rate compared to others?

97 Which part of the blood does oxygen diffuse into during gas exchange?

98 Where are red blood cells made in the body?

99 How does the shape of the red blood cell allow it to move within a narrow capillary?

100 Arteries have thicker walls than capillaries. Explain why oxygen cannot easily diffuse into an artery.

101 A person who exercises more needs to remove more of which gas?

102 What would happen to gas exchange if only one capillary lined each alveolus? Explain your answer.

103 What would happen if a trachea did not have any goblet cells?

104 What happens to the diaphragm when we breathe in?

105 What happens to a person's breathing rate when they do not move around much?

106 What are the three types of muscle in the human body?

> **Tip**
>
> Remember the function and adaptations of alveoli and the lungs when answering these questions.

What is asthma?

107 Name two triggers of an asthma attack.

108 What happens to our breathing rate when we exercise?

109 What happens if the airways get narrower?

110 What tissue is the diaphragm made of?

111 What is asthma?

112 This question is about goblet cells.

    a What is the function of goblet cells?

    b Goblet cells are found in the human trachea. Which sub-cellular structures would we expect to find in one goblet cell?

    c What happens within the trachea during an asthma attack?

    d What effect does this have on breathing? Explain why.

    e What other cell type is usually found in the trachea?

113 How do the following conditions affect breathing rate?
   a Sleep
   b Asthma attack
   c Exercise
   d Lack of blood supply surrounding the alveoli

114 What are two things that happen during an asthma attack that narrow the airways?

115 A person has an asthma attack. What would happen to this person if they did not have cartilage within their trachea?

116 How does asthma affect lung volume?

117 This question is about athletes.
   a Compare the energy requirement of an athlete with that of a spectator.
   b Suggest what foods the athlete should eat to get this energy.
   c Why does the athlete need to include proteins in their diet?
   d An athlete is diagnosed with asthma. How will this affect their breathing and breathing rate?

118 How is a ciliated epithelial cell specialised for its function?

119 Why do we release mucus into the trachea?

120 A scientist wanted to investigate if the type of substance a person is exposed to affects the width of the bronchi.
   a Write a line of inquiry for this investigation.
   b What is the independent variable?
   c What is the dependent variable?
   d Suggest two control variables.
   e The bar chart below shows the scientist's results. Which substance narrows the bronchi the most?

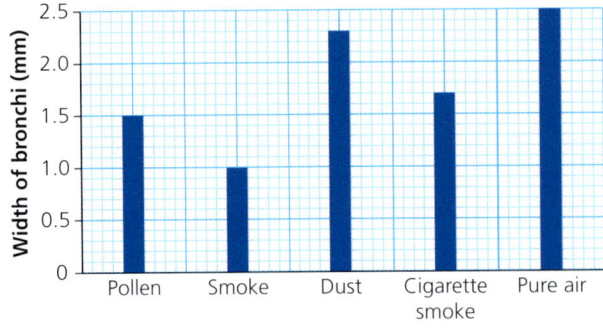

   f Looking at the bar chart, which substance (besides pure air) would be least likely to trigger an asthma attack?

121. What effect does exercise have on breathing rate?
122. In spring, lots of flowers release pollen. Why is this a problem for some people with asthma?
123. Name the two muscles involved in breathing.
124. Which type of muscle provides cushioning around organs?
125. What happens to our breathing rate when we are asleep?
126. A scientist is investigating the effect of different quantities of pollen on the width of the bronchiole.
    a. Write a line of inquiry for this investigation.
    b. What is the independent variable?
    c. What is the dependent variable?
    d. Give two control variables in this investigation.
    e. What type of graph should this scientist plot for their results?
127. What is diffusion?

How does smoking affect breathing?

128. What are four effects of smoking on the lungs?
129. How can smoking regularly eventually lead to smoker's cough?
130. Why are smokers more likely to get bronchitis?
131. Why can smoking lead to heart disease?
132. a. What is the name of the tube running from the nose and mouth to the lungs?
    b. This tube is lined with cells with hair-like structures on them. What is the function of these hair-like structures?
133. Why are the alveoli membranes so highly folded?
134. What happens to the oxygen that crosses the alveoli into the blood?
135. A person with asthma is advised not to smoke. Explain how smoking would severely affect their health.
136. This question is about a ciliated epithelial cell.
    a. Ciliated epithelial cells have many mitochondria. Suggest why.
    b. As the cilia move, which energy store is increasing?
    c. A ciliated epithelial cell is 0.0018 mm wide. How wide will it appear to be when magnified ×350?
    d. Convert your answer in part c to centimetres.
137. Why does smoking increase a person's breathing rate?
138. What effect does asthma have on the muscles surrounding the airways?

139 This question is about red blood cells.

  a What is the function of red blood cells?

  b How does smoking affect this function?

  c Mature red blood cells and bacterial cells both lack a sub-cellular structure. Name this structure.

  d If a red blood cell did not have a biconcave shape, how would this affect its function? Explain your answer.

140 The diagram shows an alveolus. Which gases do arrows A and B represent in the diagram below?

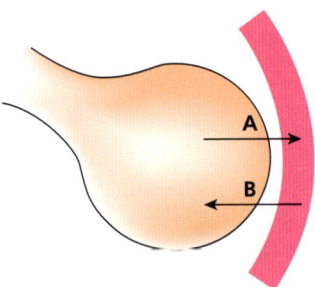

141 The diagram below shows alveoli in non-smokers and smokers. Describe how this difference can affect gas exchange.

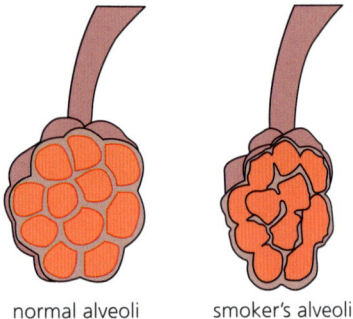

normal alveoli   smoker's alveoli

142 The table below shows the lung volumes for three different people.

Person A is a trained athlete whereas persons B and C are not.

Person	Age	Volume of air passing into the lungs (cm³)	
		At rest	During exercise
A	25	500	5000
B	25	500	4000
C	65	400	3500

  a Explain why more air is taken into the lungs during exercise.

  b Explain why the volume of air at rest is different for person B and person C.

  c Explain why the volume of air during exercise for person A is much higher than for person B.

143 The graph below shows how the number of cigarettes smoked in a day affects the breathing rate.

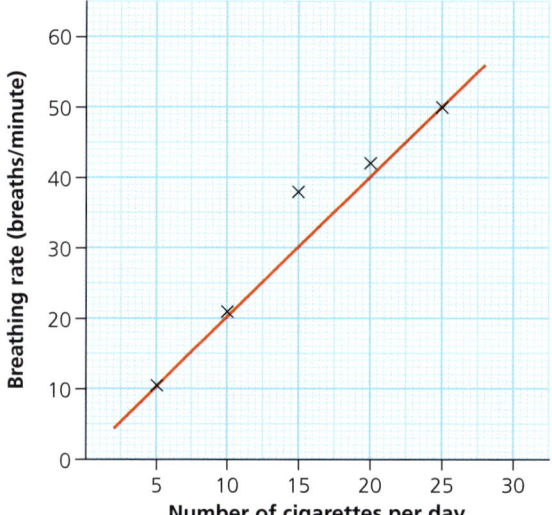

a What is the breathing rate when someone smokes 20 cigarettes a day?

b A student says that if you double the number of cigarettes you smoke in a day, you double your breathing rate. Are they correct? Explain your answer.

c A carcinogen is a chemical that causes cancer. How is tar a carcinogen?

d What effect does cigarette smoke have on ciliated epithelial cells?

e How does smoking affect the function of red blood cells?

> **Tip**
>
> Use a ruler to help you follow the lines on the graph to answer this question.

144 Regular exercise strengthens the muscles involved in breathing. Why is this a good thing?

145 A person who has asthma smokes several cigarettes a day. What changes can they make to their lifestyle to improve their breathing rate?

146 What would happen if a person did not have ciliated epithelial cells in their trachea? Explain your answer.

147 What are the little air sacs in the lungs called?

148 Which bones protect the lungs during ventilation?

149 A student says the alveoli are damaged when someone has an asthma attack. Are they right? Explain your answer.

150 What happens to lung pressure when we exhale?

151 What does smoking do to the alveoli?

152 What part of cigarettes is addictive?

153 What happens to the surface area of alveoli after many years of smoking?

154 What effect does nicotine have on blood vessels?

B5 Reproduction

B5.1 Sexual reproduction and the reproductive organs

Which cells cause sexual reproduction?

1. What is the scientific name for a child?
2. How many parents are involved in sexual reproduction?
3. What is the scientific name for sex cells?
4. What is the name of the male gamete?
5. What is the name of the female gamete?
6. Which human gamete has the ability to swim?
7. A human sperm can swim 3 mm in a minute. Calculate an average sperm's speed in metres per second.
8. This question is about gametes.
   a. What is a gamete?
   b. A sperm cell's tail is 0.04 mm long. How long will it appear to be if magnified ×380?
   c. What is the function of the nucleus?
9. Why are humans considered to be mammals in terms of reproduction?
10. Why are gametes considered specialised cells?
11. Gametes have a cell membrane. What is the function of the cell membrane?
12. What is the purpose of gametes?

> **Tip**
>
> These questions also cover 'How do mammals reproduce?'

What are the organs of the male reproductive system?

13 Name parts A–G in the diagram of the male reproductive system below.

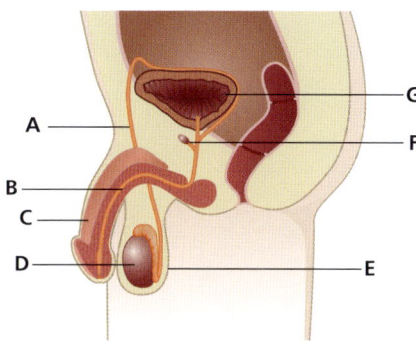

14 Which part of the diagram of the male reproductive system above is not involved in reproduction?

15 Which organ produces the sperm?

16 Which organ delivers the sperm into the vagina during sexual intercourse?

17 Which organ provides the fluid that supports the sperm on their journey?

18 The sperm start their journey at the testes and end at the penis. Write the parts of the reproductive system in the order the sperm travel through.

19 A male can choose to have a vasectomy. This is when the sperm ducts are cut or blocked. Why will a vasectomy make a male unable to father a child?

20 State the two substances that can flow down the urethra.

21 Sperm have many mitochondria. What is the purpose of mitochondria?

22 Some men can suffer from testicular cancer. In some cases, the best way to prevent the cancer from spreading and becoming a more serious problem is to remove the affected testis.

   a What is the function of the testes?

   b The testis is an organ. What is an organ?

   c Explain why a male can still produce sperm if they have had a testis removed.

   d Suggest why the male might find it harder to father a child after the operation than before.

   e Suggest why the volume of fluid produced with each ejaculation will be the same after the operation.

   f Why do sperm have a streamlined shape?

> **Tip**
> Do not include parts the sperm does not pass through.

B5 Reproduction

What are the organs of the female reproductive system?

23 Name parts A to E in the diagram of the female reproductive system below.

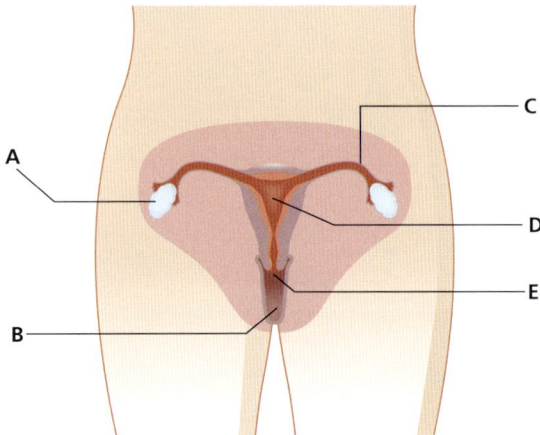

24 Which part of the female reproductive system is entered by the penis in sexual reproduction?

25 Which part is the location in which the baby will grow?

26 What is a gamete?

27 Which organ is where the female gametes develop?

28 What is the name of the female gamete?

29 Which part of the female reproductive system is a narrow passage that protects the uterus?

30 The ovum is an animal cell.

    a List the sub-cellular structures that the ovum contains.

    b Which sub-cellular structure controls what enters and leaves the cell?

    c Which sub-cellular structure is where chemical reactions occur?

    d Which sub-cellular structure is the site of respiration?

    e Name a sub-cellular structure that is not found in the ovum but would be found in a leaf cell.

31 A sperm cell is 0.05 mm long. How big will the image be when the cell is viewed under ×800 magnification?

32 Each sperm cell performs the same function. Despite this, they are not considered a tissue. Suggest why not.

B5.2 Fertilisation

How are ova adapted to their function?

33. What is a gamete?
34. Why do sperm have a streamlined shape?
35. What reaction occurs in mitochondria?
36. Why do sperm have many mitochondria?
37. What features allow the sperm to be well adapted to swimming?
38. State two adaptations of the ovum.
39. Where is the sperm's genetic information stored?
40. Which sub-cellular structure in the ovum is much larger than in the sperm?
41. How is the cell membrane of the ovum different from that of the sperm?
42. Which sub-cellular structures do plant cells have that animal cells lack?
43. What is the function of the ribosomes?
44. An ovum is 0.1mm long. How large will its image be when the cell is viewed under a ×200 magnification?

> **Tip**
> These questions also cover 'How are sperm adapted to their function?'

How does the sperm fertilise the ovum?

45. What is the scientific name for the process when the sperm and ovum join together?
46. What part of the sperm cell is transferred to the ovum when they meet?
47. Name the parts of the female reproductive system that the sperm travels through in order.
48. Where does fertilisation occur?
49. How many sperm can join with a single ovum?
50. A student says, 'The sperm swims up to the ovary to meet the egg cell and join to it.' Their answer has some mistakes. Rewrite their answer to fix the mistakes and improve the scientific language.
51. What is the name of the cell made when the sperm fertilises the ovum?
52. What is special about the amount of genetic material in the nucleus of the sperm and ovum?

53 Chlamydia is a sexually transmitted infection. It is caused by a bacterium and can damage the oviducts.

  a Suggest why this might make it difficult for a woman to conceive a child.

  b Bacteria are unicellular organisms. What does *unicellular* mean?

  c Bacteria have a flagellum. What is the purpose of the flagellum?

  d Name the other five common sub-cellular structures found in bacteria apart from the flagellum.

54 The diaphragm is a female contraceptive device that blocks the cervix. Suggest why this will prevent pregnancy.

55 The ovum is an animal cell. Name the sub-cellular structures it must contain.

56 An ovum is measured as 0.22 cm wide.

  a Convert its length into metres.

  b Calculate how big its image would be, in metres, if viewed under a ×400 microscope.

57 A doctor is looking at a sample of sperm that is from a male who has fertility issues.

  a Draw a labelled diagram of what a healthy sperm cell should look like.

  b In which organ of the male reproductive system are the sperm produced?

  c Which organ provides the fluid the sperm swim in?

  d Some of the sperm in the sample have two tails. Suggest why this might make it hard for these sperm to fertilise the ovum.

  e Some of the sperm have two heads. Suggest why this might make it hard for these sperm to fertilise the egg.

  f The doctor was observing the sperm under a ×1500 microscope. If a sperm cell was 0.07 mm long, how large would the image on the microscope be?

58 The sperm count is the number of sperm in a sample. An experiment was conducted to compare the sperm count of males who smoked different numbers of cigarettes a day. The data is in the table below.

Number of cigarettes smoked (per day)	Mean sperm count (in millions of sperm)
0	210
1–10	180
11–20	150
More than 20	130

a Write a line of inquiry for this experiment.

b What is the independent variable for this experiment?

c A non-smoking male provides a sperm sample.

   i How many sperm would we expect to count in the sample?

   ii Give a reason why the number you have given in part i might differ from the number actually counted in the sample.

d Looking at the table of results, what is the relationship between the number of cigarettes smoked and mean sperm count in millions?

e A student says, 'the independent variable is numbers so I can plot a line graph'. Explain why they are incorrect and a bar chart must be used.

f Plot a bar chart of the data provided in the table.

g Smoking also has effects on the respiratory system. Describe two effects smoking can cause in the respiratory system.

59 A doctor is looking at two patients' files before they meet them for the first time. Male A has a breathing rate of 15 breaths in 30 seconds. Male B has a breathing rate of 12 breaths in 20 seconds.

a Calculate the breathing rate of both males in breaths per minute.

b The doctor looks at the results and thinks male B might be a smoker. Give a reason why the doctor might be right and a reason why they might be wrong.

c The doctor is also looking at both males' sperm counts. Male B has a sperm count that is 75% of that of male A. Male A has a sperm count of 200 million. Calculate male B's sperm count.

d These two pieces of information make the doctor certain that male B is a smoker. Explain why the doctor might still not be correct.

B5.3 Fetal development

How does a zygote become a fetus?

60 Where does the ovum get fertilised?

61 What is the name of the cell made when the ovum is fertilised?

62 What is different about the genetic material in the ovum before it is fertilised compared to after it has been fertilised?

B5

63. A student says, 'The ovum is not a gamete after it has been fertilised.' Are they correct? Give a reason.

64. What do we call the ball of cells which implants into the lining of the uterus?

65. When the zygote is created it is one cell. After a process called cell division it forms two cells. The next division creates four cells. How many cells will there be after 10 divisions?

66. If each cell division takes 15 hours, how many days will 10 divisions take?

67. A device called an intrauterine device (IUD) can be used by females to prevent pregnancy. It makes the lining of the uterus thinner. Suggest why this stops a pregnancy occurring.

68. A student's mother is having another baby. The student says, 'My mum has been pregnant for 12 weeks. Her embryo is growing really fast.' What mistake have they made?

69. This question is about sperm.

    a State the function of the sperm.

    b Describe how it is adapted to perform that function.

How does the fetus stay alive?

70. What substances does the fetus need from the mother?

71. What substances does the fetus send to the mother?

72. What organ connects the mother and baby?

73. What process causes the oxygen to move from a high concentration in the mother's blood to the lower concentration in the fetus' blood?

74. What are the names of the male and female gamete?

75. What is the purpose of the umbilical cord?

76. What is the purpose of the amniotic fluid?

77. What is the organ that the embryo implants into at the beginning of pregnancy?

78. What do we call the first cell made from the fertilisation of the ovum?

79. Where does fertilisation occur in the female reproductive system?

80. What feature of the blood supply in the placenta increases the rate of diffusion?

81. Which sub-cellular structure controls what substances enter and leave the cell?

82. In terms of diffusion, why does the waste move from the fetus to the mother and not the other way around?

83 Name parts A to G in the diagram below of a developing fetus in the womb.

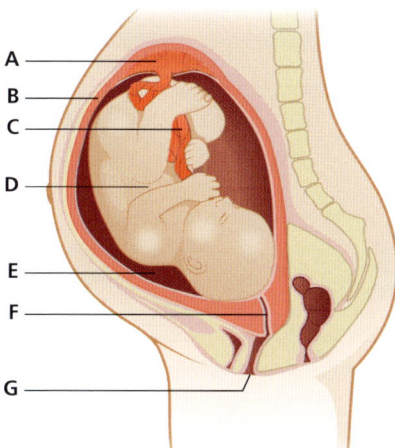

84 Some females suffer from a condition called pre-eclampsia. This condition can cause the placenta to separate from the mother's uterus. Suggest why pre-eclampsia might mean the fetus needs to be delivered earlier than it would normally be.

85 As the fetus develops, its organs begin to grow forming organ systems.

    a What is an organ?

    b How is an organ system different from an organ?

    c The lungs are part of which organ system?

    d What is the process that allows oxygen to pass from the atmosphere into the mother's lungs?

    e Explain why the carbon dioxide moves from the mother's blood into the lungs and not the other way around.

    f What is the name of the small air sacs found in the lungs?

    g How are the air sacs adapted to increase the rate oxygen moves into the blood?

    h Below are the steps of ventilation. Copy them out in the correct order to explain what happens when the mother inhales.
- Air rushes into the lungs.
- The intercostal muscles contract.
- The diaphragm contracts and flattens.
- There is a lower pressure in the lungs.
- The volume of the lungs increases.

86 A student says, 'When the fetus is growing in the uterus, it needs to absorb lots of oxygen so its lungs need to develop quickly.' Are they correct? Give a reason.

87 Some females can suffer from a condition called an ectopic pregnancy. In an ectopic pregnancy, the ovum is fertilised in the oviduct but then does not move down to the uterus.

Instead of implanting in the uterus it begins to develop into an embryo in the oviduct. This can cause symptoms in the mother such as tummy pain, vaginal bleeding and discomfort going to the toilet. The ectopic pregnancy is resolved by giving the mother medicine which stops the embryo developing and sometimes by removal of the embryo using surgery.

    a The sperm and ovum are gametes. What are gametes?

    b What is the name of the fertilised cell made when the sperm meets the ovum?

    c What adaptation of the ovum ensures only one sperm fertilises the ovum?

    d Explain why it is impossible for an ectopic pregnancy to develop into a healthy fetus.

    e Suggest why a female with damaged oviducts might have a higher risk of an ectopic pregnancy.

    f In the UK, about 1 in 90 pregnancies is ectopic. This is about 11 000 in a year. Use this information to calculate the total number of pregnancies in the UK each year.

What factors affect gestation?

88 What is the name of the developing baby in the first 8 weeks?

89 What is the name of the developing baby after 8 weeks?

90 What organ is responsible for providing the fetus with everything it needs to survive?

91 How does the fetus remove waste from its body?

92 A female has just found out she is pregnant. She currently smokes 10 cigarettes a day and goes to the pub on a Friday night to drink wine with her friends. She does not eat many vegetables. Suggest three things she could do to improve the chances of delivering a heathy baby. Give a reason for each choice.

93 The table below shows data about the development of a healthy fetus.

Age of fetus (weeks)	Approximate length of fetus (cm)
8	3
12	9
16	14
20	28
24	36
28	40

Age of fetus (weeks)	Approximate length of fetus (cm)
32	45
36	50
40	52

a What is the relationship between the age of the fetus and the length of the fetus?

b Why can we only approximate the length of the fetus?

c During which 4-week period does the fetus double in length?

d How many times longer is the fetus at 40 weeks than 8 weeks? Give your answer to three significant figures.

94 Suggest changes a pregnant female could make to their lifestyle while pregnant to increase the chance of the fetus growing as big as possible.

95 The bar chart below shows the evidence to support the impact of smoking on birth mass.

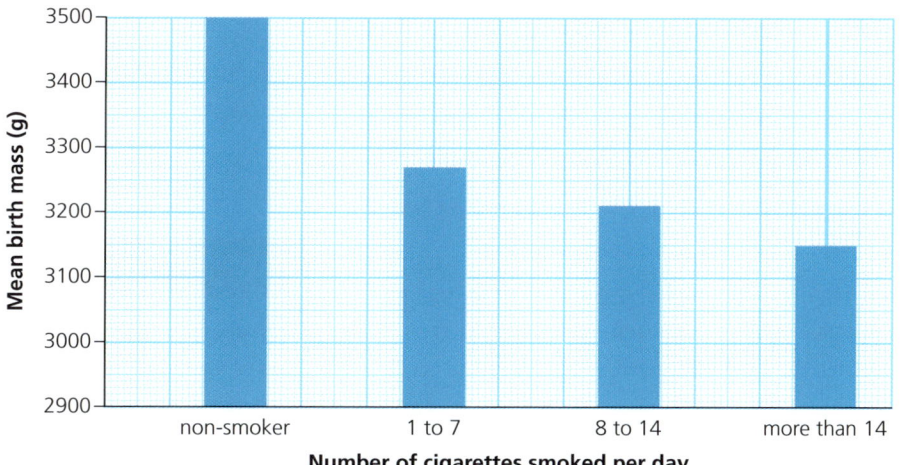

a What is the relationship between the number of cigarettes smoked per day and the mean birth mass?

b What is the mean birth mass of a child of a non-smoking mother?

c What is the mean birth mass of a child from a mother who smokes more than 14 cigarettes a day?

d Calculate the difference in mean birth mass of children born to mothers who smoke 1 to 7 cigarettes a day and those born to mothers who smoke more than 14 cigarettes a day.

e A student says, 'My mum smoked through her pregnancy with me but I was born at 3500 g. This data is rubbish!' Explain why, even though they are not lying, the data is still true.

B5 Reproduction

f What are some of the effects of smoking on the adult human body?

96 Suggest why when a female goes to the pharmacist for some medicine they are often asked if they are pregnant.

97 All mammals gestate their young inside the females' bodies. Below is a table comparing the gestation times of a range of mammals.

Mammal	Mean gestation time (days)
Mouse	19
Rhinoceros (black)	450
Sheep (domestic)	148
Whale (sperm)	480
Lion	108
Monkey (rhesus)	164
Wolf	68

a Put the mammals in order from shortest gestation time to longest gestation time.

b A human has a gestation time of 9 months. Estimate the mean gestation time in days. Show your working.

c Why do we have to put mean values in this table for each species?

d A student looks at the table and writes the following conclusion: 'The larger the mammal, the longer the gestation time.' Evaluate the student's answer.

e The student plotted a graph to display the data. The graph is shown below. Describe two errors the student has made.

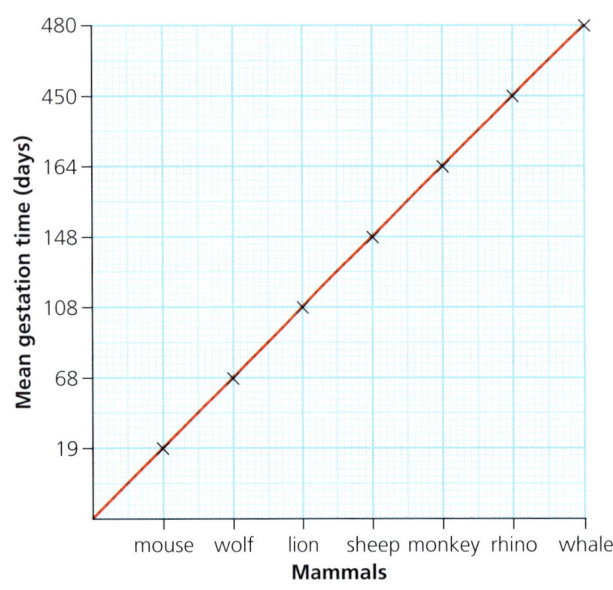

> **Tip**
>
> Evaluating needs you to explain how the student is correct and how they are wrong before making an overall judgement.

> **Tip**
>
> For help, see the 'Working scientifically' box in the 'Lines of inquiry and variables' section of your Knowledge Book and look back at the work you completed in Unit B3.

f All mammals develop a placenta during pregnancy. What is the function of the placenta?

g A zookeeper is working in the rhino enclosure. One of the rhinos is pregnant. Why is it important that the zookeeper carefully monitors the mass and types of food the pregnant rhino is eating?

h How many years is the sperm whale pregnant for? Give your answer to two decimal places.

B5.4 The menstrual cycle

How does the ovum affect the lining of the uterus?

98 What do we call the process that occurs when two gametes fuse together?

99 What does the embryo do when it enters the uterus?

100 How is the uterus adapted to support the embryo?

101 What organ develops between the uterus lining and the embryo as it grows?

102 What substances move to the embryo from the mother down the umbilical cord?

103 What substance moves from the embryo to the mother down the umbilical cord?

104 What happens to the lining of the uterus if the embryo is implanted?

105 What happens to the lining of the uterus if the egg is not fertilised?

106 Why does the body regrow the uterus lining?

107 What is the difference between a zygote, an embryo and a fetus?

108 The ovum is a gamete, which is a specialised animal cell. Give the function of the following sub-cellular structures in the ovum.

   a Nucleus

   b Cell membrane

   c Cytoplasm

   d Ribosomes

   e Mitochondria

> **Tip**
>
> These questions also cover 'How does the body prepare for a zygote?'

109 The following is a list of adaptations of specialised cells. Identify whether each one is an adaptation of the sperm, ovum, red blood cell, nerve cell or muscle cell. Some might be adaptations of more than one cell.

   a Selective cell membrane

   b Streamlined shape

   c No nucleus

   d Many mitochondria

   e Long and thin with many connections

   f Biconcave shape

   g Large cytoplasm

What is the menstrual cycle?

110 What is the common name for menstruation?

111 How long is the average pregnancy in weeks?

112 How long is the average menstrual cycle?

113 On which days of a typical menstrual cycle does menstruation occur?

114 What organ develops between the embryo and the lining of the uterus?

115 What is the function of the umbilical cord?

116 Where in the female reproductive system does fertilisation occur?

117 What happens around day 14 of the menstrual cycle?

118 A student says, 'The lining of the uterus is thick and hard so the zygote can implant into it easily.' They have made two mistakes. Rewrite the sentence with the two mistakes corrected.

119 Missing a menstruation can be one of the first signs of pregnancy. A female is 7 days past when their period is due and it has not arrived. They take a pregnancy test and it is positive. Suggest how many days they have been pregnant for. Give a reason for your answer.

120 List three adaptations of the sperm that help them to swim.

121 What is the difference between the nucleus in an ovum and the nucleus in a cell of the embryo?

122 Copy and complete the following sentences.

   a Menstruation is important for fertility because …

   b Menstruation is important for fertility, but …

   c Menstruation is important for fertility, so …

123 Some people find reproducing difficult so have a child using *in vitro* fertilisation (IVF). An ovum is taken from the female and sperm from the male. These are then combined outside of the body and the fertilised embryo is implanted into the female uterus.

　a What happens in fertilisation?

　b What organ was the ovum taken from?

　c What organ were the sperm produced in?

　d A technician observes the sperm fertilise the ovum under a ×80 microscope. The ovum is 0.1 mm wide. How wide will the image be?

　e The sperm are 0.005 mm long. What size will they appear under ×80 magnification?

　f The human eye cannot see objects that are smaller than 0.08 mm wide. Will the technician be able to see the sperm?

　g The technician swaps the magnification to ×800. Explain why this magnification is more appropriate. Give a reason for your answer.

124 The muscles that contract around the uterus are not under conscious control. What type of muscle are they?

125 Write out the following in order of size, from smallest to largest: muscle cell, nucleus, uterus, reproductive system, muscle tissue.

126 Some vegetables, such as spinach, are good sources of iron. Some people suffer from a lack of iron.

　a What is the name of the condition caused by a lack of iron?

　b Why would a person with this condition suffer from a lack of energy, tiredness and shortness of breath?

　c Suggest two ways the person could recover from the lack of iron.

127 A student is watching a documentary on obesity. In the documentary, a person with obesity has just been diagnosed with a lack of vitamin C, which comes from citrus fruits.

　a What is obesity?

　b What causes a person to become obese?

　c The student says, 'It is impossible for a person with obesity to suffer from a lack of vitamin C.' Are they correct? Give a reason for your answer.

128 Copy and complete the following sentences by filling in the gaps.

Female humans release an ovum once every _____ days. If fertilised, the embryo will develop into a _____ .

B5 Reproduction

It takes _____ months before the uterus muscles _____ and the baby is born. After the baby is born, the _____ cord connecting to the mother is cut.

129 Contraception is the term used to describe products which aim to reduce the chance of pregnancy occurring. There are two main types of contraception.
- barriers: these prevent the sperm from meeting the ovum by providing a physical barrier. Condoms are an example.
- hormonal: these chemicals flow through the bloodstream and prevent the ovum being released by the ovary. The contraceptive pill is an example.

a Human immunodeficiency virus (HIV) is a sexually transmitted infection that can be spread between people during sexual intercourse. Which type of contraception would be good at preventing pregnancy and the chance of HIV infection?

b Explain why a female taking the contraceptive pill is unlikely to get pregnant.

c The female taking the contraceptive pill has to take a pill each day. The instructions on the contraceptive pill say the following: 'Do not have unprotected sex if you have missed a pill or been sick after taking the pill.' Suggest why a female who has missed a pill or been sick might increase their chance of getting pregnant.

d The instructions go on to say: 'You can resume unprotected sex after you have had three days of successfully taking the pill.' Suggest why after a few days the chance of becoming pregnant has reduced.

e The contraceptive pill is taken orally. Which organ will be responsible for absorbing the hormones into the bloodstream?

f Name the process that will cause the hormones to move from the digestive system into the bloodstream.

B5.5 Plant reproduction

How does a plant reproduce?

130 What do we call the reproductive organs of a plant?

131 What is the name of the male plant gamete?

132 What is the name of the female plant gamete?

133 What is a gamete?

134 Where is pollen made?

135 Where is the ovum made?

136 What is the most common way for an ovum to be fertilised?

137 What is the purpose of the flower's petals?

138 What is the main species of pollinator?

139 What is nectar?

140 Why do flowers produce nectar?

141 What is the name of the entrance to the ovary that the pollen sticks to?

142 Copy and complete the table below comparing human reproduction to plant reproduction.

Human reproduction	Plant reproduction
Male gamete is sperm	
Female gamete is ovum	
Sperm are produced by the testes	
Ovum is produced by ovaries	
Sperm travel to ovum by swimming	
Ovum fertilised in oviduct	
Embryo develops in uterus	

How are seeds dispersed?

143 What is the name of the plant's male gamete?

144 What do bees drink from flowers?

145 Describe how a seed with spikes on can be dispersed by an animal.

146 What does the plant ovary develop into?

147 A student says, 'When plants self-pollinate it cannot be sexual reproduction because that needs two parents.' Are they correct? Give a reason.

148 Why do plants make delicious fruits?

149 What attracts the bees and other pollinators to the flowers of the plant?

150 Why is it important that the seeds are not digestible by the animal?

151 This question is about plant cells.

   a Which part of the plant cell is responsible for strengthening the cell?

   b Which part contains the cell sap?

> **Tip**
>
> These questions also cover 'What are fruits and seeds?'

B5

c What part contains the genetic material?

d What is the function of the mitochondria?

e What is the function of the cell membrane?

f A student says, 'The plant has a cell wall which controls what goes in and out of the cell.' Are they correct? Give a reason.

g What sub-cellular structure is where photosynthesis occurs?

How are seeds dispersed? *continued*

152 What does *dispersed* mean?

153 Why do plants try to spread their seeds far away?

154 What do we call the process where a seed develops into a new plant?

155 List three ways animals help to disperse seeds.

156 How are seeds adapted for being dispersed by the wind?

157 Some seeds get stuck in a dog's fur when it runs through long grass. Explain how this helps the grass to disperse its seeds.

158 What happens when a plant uses explosion methods to disperse seeds?

159 A pig eats an apple off the ground. Explain how this could lead to a new apple tree growing at the other end of the field.

160 A student says, 'Oak trees cannot use squirrels to disperse acorns because squirrels eat acorns.' Are they correct? Give a reason for your answer.

161 A student is walking past a river. They notice some round seed pods. As they touch them, the seed pods burst open. Which type of seed dispersal method is being used?

162 A student is blowing a dandelion clock in their grandparents' garden. 'Stop that! You will give me more weeds to dig up,' their grandfather calls. Explain why the student's action will lead to more dandelion weeds growing.

163 The sycamore tree produces seeds that have a large wing. As they fall, they spin and slow down. What seed dispersal method is this?

164 A dog eats a seedless grape. Will it help the germination of a new grape vine? Give a reason for your answer.

165 What do we call the process where a seed begins to develop into a new plant?

166 Name the sub-cellular structures that are in plant cells but not animal cells.

167 Which sub-cellular structure allows photosynthesis to occur?

168 When a seed germinates it makes shoots and roots. Name parts A to D of the plant root hair cell in the diagram below.

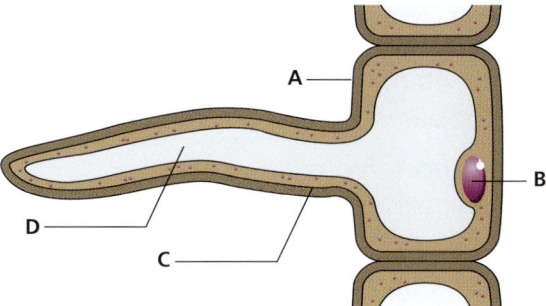

169 Explain why a root hair cell has no chloroplasts.

170 A coconut is floating on the Pacific Ocean.
   a What method of seed dispersal is being used?
   b Name the force that prevents the coconut from sinking.
   c Why can we say that the resultant force acting vertically is zero?
   d The coconut travels 200 m in 30 minutes. Calculate the speed in metres per second.

> **Tip**
>
> For help with the speed equation, see Unit P2 of your Knowledge Book. Remember there are 60 seconds in 1 minute.

171 Plant seeds are eaten by animals because they are full of fats and oils.
   a What is the function of fats and oils as part of a healthy diet?
   b What is the function of enzymes in the digestive system?
   c What are enzymes made of?
   d In what organ are the digested products of the seeds absorbed into the blood?
   e How is the organ named in part d adapted to increase the speed of absorption?
   f The nutrients move into the blood through a process called diffusion. What is diffusion?

172 A student wants to observe some onion cells under the microscope. They prepare a slide and take it to their desk where a microscope is resting.
   a Describe the steps the student must take to view a clear magnified image of the onion cells.
   b The student says, 'The onion cells look bigger because we have zoomed in on them.' Are they correct? Give a reason for your answer.
   c The onion cell is 0.25 mm long. When set up, the microscope has a total magnification of ×750. Calculate the length of the onion cell in the image.

B5 Reproduction

d When magnified, the student can notice two structures: a thick line around the outside of each cell and a dark dot. What are the names of the two sub-cellular structures the student has seen?

e Explain why it is not possible for the student to see any other sub-cellular structures with their microscope.

f The root hair cell of an onion is 1.5 mm long. How many times longer is the root hair cell compared to the onion cell?

g When an onion is planted, it grows a shoot and roots. Explain why the shoot is green but the roots are not.

h When an onion plant flowers, it produces small white petals and lots of nectar. Explain how the nectar helps the flower to get pollinated.

173 This question is about leaves.

  a What is the name of the green pigment in leaves and what is its purpose?

  b A diagram of a plant cell is shown below. Name the sub-cellular structures A to H.

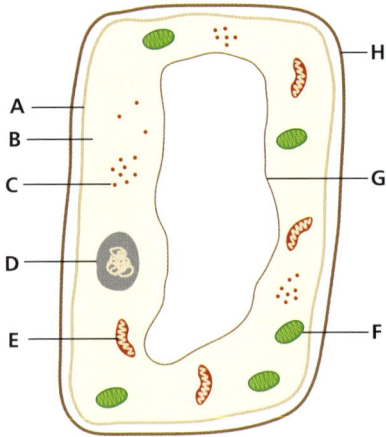

  c Which letters correspond to the sub-cellular structures which are found in animal and plant cells?

  d Which letter corresponds to the sub-cellular structure that contains the sap?

  e Which letter corresponds to the sub-cellular structure which provides strength and protects the plant cell?

  f Which letter corresponds to the sub-cellular structure which makes proteins?

  g When humans eat vegetables, they are consuming plant cells. What do we call the indigestible material which helps to bulk up the food to pass easily through the digestive system?

  h In the digestive system, what is the name of the proteins that speed up chemical reactions?

C4 Chemical reactions

C4.1 Reactions, conventions and signs a reaction has occurred

What goes in a chemical equation?

1. What is a chemical formula?
2. Why do chemical equations use '→' and not '='?
3. What do we call the starting chemicals in a chemical reaction?
4. What do we call the end chemicals in a chemical reaction?
5. Below is a word equation for the reaction between limestone and acid.

   calcium carbonate + hydrochloric acid → calcium chloride + water + carbon dioxide

   a. Name a product in this reaction.
   b. Name a reactant in this reaction.
   c. How many products are there in this reaction?
   d. How many reactants are there in this reaction?

6. Below is a word equation for the reaction between lithium and oxygen.

   lithium + oxygen → lithium oxide

   a. Name a product in this reaction.
   b. Name a reactant in this reaction.
   c. How many products are there in this reaction?
   d. How many reactants are there in this reaction?

7. Below is a word equation for the reaction between magnesium and copper nitrate.

   magnesium + copper nitrate → copper + magnesium nitrate

   a. Name a product in this reaction.
   b. Name a reactant in this reaction.
   c. How many products are there in this reaction?
   d. How many reactants are there in this reaction?

8. Below are some descriptions of chemical reactions. Turn the sentences into word equations.

   a. When methane gas burns in the presence of oxygen, the products are water and carbon dioxide.

> **Tip**
> These questions also cover 'What is a chemical reaction?'

b Copper oxide reacts with sulfuric acid making copper sulfate and water.

c Magnesium when placed in hydrochloric acid turns into magnesium chloride and hydrogen gas.

d Potassium reacts with oxygen in the air forming potassium oxide.

9 What happens in a chemical reaction?

10 Describe how particles move in:

a solids **b** liquids **c** gases.

11 What is the melting point of a substance?

12 What is the boiling point of a substance?

When are changes not a chemical reaction?

13 What is the difference between a chemical reaction and a physical change?

14 Write down the state symbols for the following four states of matter.

a Solid
b Gas
c Aqueous solution
d Liquid

15 We can represent physical state changes in equations. For example, water (ice) melting can be represented as:

$H_2O(s) \rightarrow H_2O(l)$

Write equations showing the following state changes.

a Water boiling
b Methane (CH_4) condensing
c Mercury (Hg) melting
d Oxygen boiling
e Zinc melting

16 Explain why none of the equations in Question 15 are chemical reactions.

17 Explain why chocolate melting is not a chemical reaction.

18 A student takes a piece of paper and tears it into four parts. Has a chemical reaction taken place? Explain your answer.

19 Gallium (Ga) has a melting point of 30°C and a boiling point of 2400°C.

a What state is gallium at room temperature (20°C)?

b What happens to gallium when it is heated above 30°C?

c Write a symbol equation with state symbols representing gallium melting.

d Explain why this equation does not represent a chemical reaction.

20 Bromine (Br_2) has a melting point of −7°C and a boiling point of 59°C.

    a What state is bromine at room temperature (20°C)?

    b To what temperature must bromine be heated to turn it into a gas?

    c Write a symbol equation with state symbols representing bromine boiling.

    d Explain why this equation does not represent a chemical reaction.

    e Bromine dissolves in water producing a solution of bromine water. Write a symbol equation with state symbols representing bromine becoming a solution.

    f Explain why this equation does not represent a chemical reaction.

21 Alex and Isabel were given a mixture of iron filings and some sugar.

    a Alex separated the mixture using a magnet, which attracted the iron filings and not the sugar. Explain why this is not a chemical reaction.

    b Isabel added water to the iron filings and sugar to dissolve the sugar. They then filtered the iron filings, which were insoluble. Explain why this is not a chemical reaction.

22 A student makes a glass of orange squash by adding orange squash and water. Has a chemical reaction taken place? Explain your answer.

23 A student watches an ice cube melt and says, 'A new substance, water, is formed therefore a chemical reaction has taken place.' Explain why they are wrong.

How can we tell if a chemical reaction has happened?

24 Name five signs that a chemical reaction has taken place.

25 What is *effervescence*?

26 What is *precipitation*?

27 Below are some sentences describing chemical reactions. Write a word equation for each one.

    a Magnesium is a silvery metal. When it is heated in oxygen, it makes a bright light. The product is a white powder called magnesium oxide.

> **Tip**
>
> Only the names of chemicals (not their colours or states) are put in word equations.

b When grey iron filings are mixed with yellow sulfur, a mixture is formed that can be separated with a magnet. However, after heating, the mixture glows and black iron sulfide is formed, which is a new compound.

c When grey calcium metal is added to a test tube containing hydrochloric acid, bubbles of hydrogen are formed rapidly, the solution gets very hot and a solution of calcium chloride is left in the test tube.

d When colourless solutions of lead nitrate and potassium iodide are mixed together, a yellow solid of lead iodide forms in the solution of potassium nitrate.

e When silver sodium is added to water, it fizzes and burns. It forms sodium hydroxide solution and hydrogen gas.

f Inside car exhausts, two poisonous gases, carbon monoxide and nitrogen monoxide, are converted into two less harmful gases, carbon dioxide and nitrogen. This is facilitated by the hot platinum catalyst in the exhaust box.

28 For each of the reactions described in Question 27, name the signs that a chemical reaction has taken place.

29 What are the three types of subatomic particle found inside atoms?

30 Which subatomic particles are found in the nucleus of atoms?

31 Which subatomic particles are found in shells around the nucleus?

32 A student dissolves some salt in water. They can no longer see the salt. They say that a chemical reaction must have taken place. Explain why they are wrong.

33 A student strikes a match.

   a What will they observe?

   b What will they be able to feel near the match?

   c The student blows the match out. How will the matchstick have changed?

   d Give three ways the student can tell that a chemical reaction has happened.

34 A student boils some water in a kettle. They say that because bubbles of steam are formed and the kettle gets hot that a chemical reaction has taken place. Explain why they are wrong.

35 A student is injured playing rugby. Their teacher gives them an 'instant' icepack. To activate it, a small bag inside the big bag is broken to mix two chemicals together, which makes the whole bag cold. Explain how you know that this must be a chemical reaction.

> **Tip**
>
> Observe means 'see'.

C4.2 Combustion, thermal decomposition, oxidation and displacement

What are combustion equations?

36. What type of energy is stored in fuels such as petrol, diesel and methane?
37. What is the scientific name for burning reactions?
38. What gas in the air is essential for combustion reactions?
39. What are the three sides of the fire triangle?
40. What three factors affect the pressure of a gas?
41. Butane (C_4H_{10}) is a fuel often used in barbeques.
    a. Write a word equation for the combustion of butane.
    b. Write a symbol equation for the combustion of butane.
    c. How many types of atom are in butane?
    d. How many atoms are there in total in butane?
42. Methanol (CH_4O) is used in spirit burners.
    a. Write a word equation for the combustion of methanol.
    b. Write a symbol equation for the combustion of methanol.
    c. How many elements are there in methanol?
    d. How many atoms are there in total in methanol?
43. What is an element?
44. What is a compound?
45. Copy and complete the following sentences.
    a. A fire can be extinguished by removing oxygen because …
    b. A fire can be extinguished by removing oxygen, but …
    c. A fire can be extinguished by removing oxygen, so …
46. Fire breaks are gaps in forests where trees are cut down. They help to stop forest fires spreading. Use the fire triangle to explain how fire breaks help to stop forest fires spreading.
47. How many different types of atom are there?
48. Where are all the different types of atom listed?
49. How do different atoms differ from each other?
50. Angela and Mario are watching a Bunsen burner flame. Angela says that a chemical reaction is taking place because there is heat given off. Mario says that there is no chemical reaction because they cannot see any new substances. Explain which student is correct and explain to the incorrect student why they are wrong.

> **Tip**
>
> These questions also cover 'What is combustion?"

C4

51 A student is trying to start a bonfire with some kindling wood. The wood is just catching fire, but the combustion is slow. The student blows on the flame gently and the rate of combustion increases. Explain why the combustion reaction proceeds quicker when the student blows on the flame.

What is thermal decomposition?

52 What happens in thermal decomposition reactions?

53 Copy and complete the following thermal decomposition of metal carbonates word equations.

  a sodium carbonate → sodium oxide +

  b potassium carbonate →   + carbon dioxide

  c barium carbonate →   +

  d   → iron oxide + carbon dioxide

  e   → calcium oxide +

54 Copy and complete the following table of the state symbols we use in symbol equations.

State	Symbol
	(s)
	(l)
	(g)
	(aq)

55 Copy and complete the following thermal decomposition of metal carbonate symbol equations.

  a $CaCO_3(s)$ → $CaO(s)$ +

  b $MgCO_3(s)$ →   + $CO_2(g)$

  c   → $BaO(s)$ + $CO_2(g)$

  d $ZnCO_3(s)$ →   +

56 Limewater is the chemical test for carbon dioxide. Describe what happens to limewater when carbon dioxide is bubbled through it.

57 Limewater is a solution of calcium hydroxide and water.

  a Which chemical is the solvent in a calcium hydroxide solution?

  b Which chemical is the solute in a calcium hydroxide solution?

58 Describe how a solute can be separated from a solution.

59 Metal hydroxides are not the only chemicals which undergo thermal decomposition. Another example is hydrogen peroxide (H_2O_2), a colourless liquid which decomposes to form liquid water (H_2O) and bubbles of oxygen (O_2) gas when it is heated.

    a Write a word equation of the decomposition of hydrogen peroxide.

    b Write a symbol equation of the decomposition of hydrogen peroxide.

    c Add state symbols to your symbol equation.

    d How can you tell from these equations that this is a thermal decomposition reaction?

    e Predict the signs of a chemical reaction taking place that you would observe if a beaker of hydrogen peroxide was heated.

60 Phosphorus has an atomic number of 15 and an atomic mass of 31.

    a How many protons, neutrons and electrons does phosphorus have?

    b Phosphorus is in period 3, group 5 of the Periodic Table. How are its electrons arranged?

61 A student is watching a wood fire. They say that this is a thermal decomposition reaction because the wood is being turned into carbon dioxide and water and there is heat. Explain why they are incorrect.

What is oxidation?

62 What is an oxidation reaction?

63 What reactive gas makes up approximately 20% of the Earth's atmosphere?

64 Copy and complete the following oxidation word equations.

    a potassium + oxygen →

    b lead + oxygen →

    c aluminium + oxygen →

    d   + oxygen → gallium oxide

    e manganese +   → manganese oxide

    f   +   → caesium oxide

65 Rust is the common name for red iron oxide and rusting is the common name for the process by which iron objects slowly oxidise in the air.

    a What chemical in air makes iron rust?

    b Although rusting is generally slow, what signs are there that a chemical reaction has taken place?

c Write a word equation for the process of rusting.

d Iron oxide has the formula Fe_2O_3. Write a symbol equation for the rusting of iron.

e Salt water can act as a catalyst to make rusting happen faster. Why is salt water not included in the chemical equation for rusting?

f Which chemicals in the rusting equation are elements?

g Which chemicals in the rusting equation are compounds?

h Use the information above and your own knowledge of rust to explain how compounds are different to the elements that form them.

What is a displacement reaction?

66 What is a displacement reaction?

67 Copy and complete the following displacement word equations.

    a lithium oxide + potassium → potassium oxide + _____

    b aluminium oxide + magnesium → _____ + aluminium

    c nickel oxide + titanium → _____ + _____

    d tin oxide + carbon → _____ + _____

    e _____ + _____ → strontium oxide + zinc

68 Why is '→' used in the equations above and not '='?

69 When magnesium oxide and carbon are heated there is no reaction. Suggest why a displacement reaction does not occur between these two chemicals.

70 Around 3300 BC, the Bronze Age began when humans began smelting (extracting) metals such as copper and tin and making metal tools for the first time. It is likely that the first humans to discover copper smelting did so by accident; building a wood fire on some green rocks (copper compounds) and in the morning finding pink copper flecks in the ashes. Copper had been displaced by carbon in the wood ash.

    a What type of reaction extracts copper from copper compounds using carbon?

    b Which is the more reactive element, copper or carbon?

    c Explain your answer to part b.

    d Write a word equation for the reaction of copper oxide and carbon.

    e Write a symbol equation for the reaction of copper oxide and carbon. Copper oxide has the formula CuO.

f What signs of a chemical reaction might you expect to see when this reaction takes place?

g Explain how you could test the gas produced and show that it was carbon dioxide.

h A rock contains a mixture of soluble copper chloride and insoluble copper oxide. The rock is ground into a fine powder. Explain how the two substances could be separated from each other.

71 How many different elements are there?

72 a What are physical properties of chemicals?

b Give three examples of physical properties.

73 For each of the equations below, identify whether it is a combustion reaction, an oxidation reaction, a thermal decomposition or a displacement reaction.

a iron carbonate → iron oxide + carbon dioxide

b tin + oxygen → tin oxide

c propane + oxygen → carbon dioxide + water

d barium oxide + lithium → lithium oxide + barium

e hexane → butane + ethene

f carbon + oxygen → carbon dioxide

g zinc chloride + magnesium → magnesium chloride + zinc

h oxygen + propanol → water + carbon dioxide

74 For each example in Question 73, explain how you identified that it was that type of reaction.

C4.3 Conservation of mass

How do we calculate masses of reactants or products?

75 What is the law of conservation of mass?

76 Sodium and chlorine react together to form sodium chloride in the following equation.

sodium + chlorine → sodium chloride

a If 4.6 g of sodium reacts with 7.1 g of chlorine gas, what mass of sodium chloride will be produced?

b If 23 g of sodium reacts with 35.5 g of chlorine gas, what mass of sodium chloride will be produced?

c If 39 g of sodium chloride is produced using 15.3 g of sodium, what mass of chlorine gas has it reacted with?

Tip

These questions also cover 'What is conservation of mass?'

Tip

For help, see the Worked example in Topic C4.3 of your Knowledge Book.

C4

77 Hydrogen is reacted with oxygen to produce water.

    a Write a word equation for this reaction.

    b What mass of water is produced by the reaction of 4g of hydrogen with 32g of oxygen?

    c What mass of water is produced by the reaction of 1g of hydrogen with 8g of oxygen?

    d How much hydrogen is required to produce 1.8g of water from 1.6g of oxygen?

    e How much oxygen is required to produce 54g of water from 6g of hydrogen?

78 Potassium has an atomic mass of 39 and an atomic number of 19. How many protons, neutrons and electrons does potassium have?

79 The thermite reaction is a displacement reaction that produces molten iron. In the thermite reaction, aluminium displaces iron from iron oxide, producing iron and aluminium oxide. Identify the missing masses of reactants or products.

aluminium	+	iron oxide	→	iron	+	aluminium oxide
16 g	+	5.4 g	→	10.2 g	+	a
8 g	+	b	→	5.1 g	+	5.6 g
c	+	7.7 g	→	14.6 g	+	16 g

80 What technique is used to separate different colour inks?

81 Sketch a diagram of the equipment required for a filtration of an insoluble substance from a solution. Label the filtrate and the residue.

82 A student dissolves 6.8g of sugar in 50g of water.

    a What is the mass of sugar solution at the end?

    b This solution is not saturated. What does this mean?

    c Explain why this is not a chemical reaction.

    d How could the student get the 6.8g of sugar back?

When might it appear that mass is not conserved in a reaction?

83 Draw the particle arrangement for a solid, a liquid and a gas.

84 When would reactions not appear to obey the law of conservation of mass?

85 When would reactions appear to gain mass?

86 When would reactions appear to lose mass?

87 A student gently heats a test tube containing green copper carbonate, which turns into black copper oxide and carbon dioxide.

    a Write a word equation for this reaction.

    b What sign is there that a chemical reaction has taken place?

    c What type of reaction is this?

    d The student weighs the test tube before and after the reaction. Explain what will happen to the measurement of mass after the reaction.

88 A class complete a practical to react magnesium metal with oxygen from the air. They complete the reaction in crucibles with lids to try to stop magnesium oxide from escaping. They measure the mass of magnesium they start with and the mass of magnesium oxide they have at the end. Below is a table of their results.

Group	Mass of magnesium (g)	Mass of magnesium oxide (g)
A	1.3	2.1
B	1.9	3.2
C	2.8	4.7
D	3.5	5.7
E	4.2	6.9
F	4.9	7.2
G	5.5	9.2

    a Draw a graph of the class data.

    b Draw a circle around the anomalous point.

    c Draw a line of best fit through the rest of the data.

    d Suggest a reason why the anomalous point did not fit the trend.

89 A student sets up an experiment to see what happens to an iron nail left in some water. They measure the mass of the nail at the start, and it has a mass of 2.45 g. At the end of one week, they come back and notice that the nail is rusty. It now has a mass of 2.58 g.

    a Explain why the nail has increased in mass.

    b Write a word equation for the reaction.

    c What type of reaction is this?

d The student notices there is more solid rust that has fallen off the nail in the water. Suggest how they could separate this solid rust from the water so they can measure the mass of this too.

90 Copy and complete the following sentences.

    a In chemical reactions, the mass of products is equal to the mass of reactants because ...

    b In chemical reactions, the mass of products is equal to the mass of reactants, but ...

    c In chemical reactions, the mass of products is equal to the mass of reactants, so ...

How do we balance symbol equations?

91 Draw see-saw bubble diagrams for the following symbol equations and state whether they are balanced or not.

    a $Na + F_2 \rightarrow NaF$

    b $HCl + NaOH \rightarrow NaCl + H_2O$

    c $CuCO_3 \rightarrow CuO + CO_2$

    d $Fe_2O_3 + Al \rightarrow Al_2O_3 + Fe$

    e $CuSO_4 + Mg \rightarrow MgSO_4 + Cu$

92 Balance the following symbol equations. Ensure you write the full balanced equation at the end of your answer.

    a $H_2 + Br_2 \rightarrow HBr$

    b $Mg + HCl \rightarrow MgCl_2 + H_2$

    c $Na + O_2 \rightarrow Na_2O$

    d $Cl_2 + Al \rightarrow AlCl_3$

    e $CaCO_3 + HCl \rightarrow CaCl_2 + H_2O + CO_2$

    f $C_6H_{14} \rightarrow C_2H_6 + C_2H_4$

    g $Na + H_2O \rightarrow NaOH + H_2$

93 What is a balanced symbol equation?

94 Where are the reactants in a chemical equation?

95 Where are the products in a chemical equation?

96 Why is the chromatography line drawn in pencil?

97 What is diffusion?

98 What are the names of the following chemicals?

    a H_2O **c** CaF_2 **e** $NaBr$

    b PbO **d** $AlCl_3$ **f** CH_4

> **Tip**
>
> Remember that both the number and types of atom on each side must be the same, not just the total number of atoms.

> **Tip**
>
> For help, see the Worked example in Topic C4.3 of your Knowledge Book.

99 What are the formulae of the following chemicals?
   a Carbon dioxide
   b Carbon monoxide
   c Nitrogen dioxide
   d Sulfur trioxide

C4.4 Acids, alkalis and pH

What substances are acids and alkalis?

100 What scale is used to measure the strengths of acids and alkalis?

101 What name is given to a substance that is neither an acid nor an alkali?

102 Hydrochloric acid is a strong laboratory acid with the formula $HCl(aq)$.
   a How many different atoms are in the formula of hydrochloric acid?
   b Is hydrochloric acid an element or a compound? Explain your answer.
   c What does the state symbol '(aq)' mean?
   d What pH would you expect hydrochloric acid to have?
   e What precaution should be taken when using strong corrosive acids in the laboratory?

103 Sodium hydroxide is a strong laboratory alkali with the formula $NaOH(aq)$.
   a How many different atoms are in the formula of sodium hydroxide?
   b Is sodium hydroxide an element or a compound? Explain your answer.
   c What does the state symbol '(aq)' mean?
   d What pH would you expect sodium hydroxide to have?
   e What precaution should be taken when using strong corrosive alkalis in the laboratory?

104 Water is a neutral substance with the formula $H_2O(l)$.
   a How many different types of atom are in the formula of water?
   b Is water an element or a compound? Explain your answer.
   c What does the state symbol '(l)' mean?
   d What pH would you expect water to have?
   e Are there any hazards associated with water?

> **Tip**
> These questions also cover 'What are acids and alkalis?'

105 A student measures the pH of several unknown solutions using an electronic digital pH meter. Below are their results. Copy and complete the table, stating whether each solution is a strong acid, weak acid, strong alkali, weak alkali or neutral substance.

Solution	pH	Type of substance
A	1.3	
B	8.7	
C	13.8	
D	6.5	
E	7.0	
F	0.3	

106 Give an example of a household alkali.

107 Give an example of a household acid.

108 Give an example of a neutral household substance.

109 Explain why safety precautions must be taken when handling laboratory acids such as hydrochloric acid and sulfuric acid, but not when handling household acids such as fizzy drinks.

What are bases?

110 What are bases?

111 What are alkalis?

112 Limestone is made from calcium carbonate ($CaCO_3$), which is an example of an insoluble base.

  a How many different types of atom are in calcium carbonate?

  b How many atoms in total are in calcium carbonate?

  c Some limestone chips are added to water. What would you expect to happen?

  d An excess (too much) of limestone chips is added to hydrochloric acid. What will happen to the pH of the solution?

113 Potassium hydroxide (KOH) is an example of a soluble base.

  a How many different types of atom are in potassium hydroxide?

  b How many atoms in total are in potassium hydroxide?

  c Some potassium hydroxide is added to water. What would you expect to happen?

  d An excess of potassium hydroxide is added to sulfuric acid. What will happen to the pH of the solution?

114 Indigestion is when excess stomach acid (HCl) causes discomfort. Indigestion remedies contain insoluble bases such as sodium hydrogen carbonate ($NaHCO_3$). Suggest why insoluble bases are used to treat indigestion rather than soluble alkalis.

What are indicators?

115 What is universal indicator?

116 What colour is universal indicator in acids?

117 What colour is universal indicator in alkalis?

118 What colour is universal indicator in neutral solutions?

119 A student measures the pH of some laboratory chemicals using universal indicator and they observe the following colour changes. What types of chemical are these solutions?

  a Universal indicator turns red.

  b Universal indicator turns blue.

  c Universal indicator turns green.

  d Universal indicator turns yellow.

120 Copy and complete the table below with missing pH values and colours.

Substance	Approximate pH	Colour of universal indicator
Coffee	5	
Bleach		Purple
Soap	9	
Vinegar	3	
Milk		Yellow
Salt water		Green

121 What pH are strong acids?

122 What pH are weak alkalis?

123 What pH are weak acids?

124 What pH are neutral substances?

125 What pH are strong alkalis?

126 A student measures the pH of some solutions using a digital pH probe. What types of solution would have the following pH measurements?

  a pH 7.0   c pH 12.2   e pH 7.3

  b pH 1.3   d pH 6.8   f pH 0.2

127 A student sets up an investigation into a reaction between sulfuric acid and magnesium carbonate. The following are the steps of their investigation and their observations.
- They poured some sulfuric acid in a beaker and added some universal indicator.
- They then added a spatula of magnesium carbonate. The reaction fizzed and the universal indicator turned orange.
- They added two more spatulas of magnesium carbonate. There was more fizzing and universal indicator turned green.
- When they added more spatulas of magnesium carbonate, there was no further fizzing and no further colour change. The magnesium carbonate solid floated in the solution.

a What colour will the universal indicator turn in sulfuric acid?

b What signs were there that a chemical reaction had taken place when magnesium carbonate was added to sulfuric acid?

c What pH had the solution become after the first spatula of magnesium carbonate was added?

d What pH had the solution become at the end of the investigation?

e State whether magnesium carbonate is a soluble alkali or an insoluble base. Explain your answer using the student's observations.

C4.5 Reactions of acids with metals and alkalis

How do we name salts?

128 Copy and complete the following neutralisation word equations.

- **a** hydrochloric acid + copper hydroxide → copper chloride +
- **b** sulfuric acid + barium hydroxide → + water
- **c** nitric acid + lithium hydroxide → + water
- **d** ___ + magnesium hydroxide → magnesium sulfate + water
- **e** sulfuric acid + ___ → sodium sulfate +
- **f** ___ + ___ → strontium chloride + water
- **g** nitric acid + rubidium hydroxide → +

129 Insoluble bases such as metal oxides can be used to neutralise acids to produce salts and water. Identify which acids react with these metal oxides to produce the following salts and write out the whole equation.

> **Tip**
> These questions also cover 'What is a neutralisation reaction?'

a	+	copper oxide	→	copper sulfate	+	water
b	+	zinc oxide	→	zinc nitrate	+	water
c	+	beryllium oxide	→	beryllium sulfate	+	water
d	+	iron oxide	→	iron chloride	+	water

130 A student wishes to make a sample of copper chloride ($CuCl_2$) salt. They react together some hydrochloric acid (HCl) and copper oxide (CuO).

    a Write a word equation for this reaction.

    b Write a balanced symbol equation for this reaction.

    c What pH are the products?

    d The student added an excess of copper oxide. State how the excess insoluble copper oxide could be removed.

    e The copper chloride filtrate is an aqueous solution. Describe how dry copper chloride salt could be produced from this solution.

> **Tip**
>
> Do not forget to write both products of this reaction.

How do metals and acids react?

131 Write the general equation for the reaction between metals and acids.

132 Hydrogen is a very flammable gas that reacts violently with oxygen to form water.

    a Write a word equation for this reaction.

    b Write a balanced symbol equation for this reaction.

133 The following symbol equation represents the reaction between a metal and an acid.

$Co(s) + HCl(aq) \rightarrow CoCl_2(aq) + H_2(g)$

    a Copy out the equation and balance it correctly.

    b What state is the cobalt chloride in?

    c What state is the hydrogen in?

    d Explain how the hydrogen could be tested to confirm that it is hydrogen.

134 Copy and complete the following word equations for the reactions of metals and acids.

a	zinc	+	sulfuric acid	→	zinc sulfate	+	
b	iron	+	nitric acid	→		+	hydrogen
c		+	hydrochloric acid	→	magnesium chloride	+	
d		+		→	barium nitrate	+	
e		+		→	vanadium chloride	+	

135 A student placed a piece of magnesium in some sulfuric acid.

  a What sign of a chemical reaction would you expect to observe?

  b Write a word equation for this reaction.

  c The salt produced is dissolved in water and has the formula $MgSO_4$. Write a symbol equation for this reaction, including state symbols.

136 A student takes an unknown piece of grey solid and adds it to a test tube containing some nitric acid. The solution is effervescent. The student is unsure if this gas is carbon dioxide or hydrogen. Explain how they could test for both gases to determine the identity of the gas.

137 For each of the phenomena below, explain whether it is a chemical reaction or a physical change. For each example, explain how you know.

  a Candle wax is melted.

  b Candle wax is combusted.

  c A bubble of methane rises into the air.

  d The bubble of methane is set on fire by the teacher.

  e A nail is hammered into some wood.

  f The nail goes rusty.

  g The wood is burned.

  h An egg is cracked into a bowl.

  i An egg is cooked in a frying pan.

138 A student takes some dry ice ($CO_2(s)$) from a freezer and watches it sublimate into a gas. (Sublimate means turn straight from a solid to a gas without becoming a liquid.)

  a Explain why this is not a chemical reaction.

  b Write a symbol equation including state symbols to show the physical change that has taken place.

  c Carbon dioxide gas can be dissolved in water to make fizzy drinks. Some of it reacts with water to make a weak acid called carbonic acid. Is this process a chemical reaction or a physical change?

139 For each of the reactions below, state whether it is a combustion, oxidation, displacement, thermal decomposition or neutralisation reaction.

  a magnesium oxide + hydrochloric acid → magnesium chloride + water

  b gallium carbonate → gallium oxide + carbon dioxide

  c silver nitrate + iron → iron nitrate + silver

d decane + oxygen → carbon dioxide + water

e decane → heptane + propene

f vanadium + oxygen → vanadium oxide

g strontium carbonate → strontium oxide + carbon dioxide

h sodium hydroxide + sulfuric acid → sodium sulfate + water

140 Define each of the following types of reaction.

  a Combustion reactions

  b Oxidation reactions

  c Thermal decomposition reactions

  d Displacement reactions

  e Neutralisation reactions

141 For each of the reactions below, balance the equation and identify the type of reaction that it is.

  a $CH_3OH(l) + O_2(g) \rightarrow CO_2(g) + H_2O(g)$

  b $Li(s) + O_2(g) \rightarrow Li_2O(s)$

  c $Fe_2O_3(s) + Al(s) \rightarrow Fe(l) + Al_2O_3(s)$

  d $Na_2CO_3(s) \rightarrow Na_2O(s) + CO_2(g)$

  e $H_2SO_4(aq) + NaOH(aq) \rightarrow Na_2SO_4(aq) + H_2O(l)$

142 Name the three types of laboratory acid and give their formulae.

143 Laboratory acids are all strong acids.

  a What pH do they have?

  b What colour will they make universal indicator turn?

144 An unlabelled solution from the laboratory is tested with universal indicator, which turns purple.

  a What pH is this substance?

  b What type of substance is this?

  c Give the name of one possible substance this may be.

145 What is the difference between an alkali and a base?

146 Magnesium oxide is a base. It is added to some nitric acid to neutralise it.

  a What pH will the acid have at the start?

  b What pH will the reaction have at the end?

  c What is the name of the salt produced in this reaction?

  d What is the other product made in this reaction?

  e Write a word equation for this reaction.

147 A piece of zinc is added to some sulfuric acid.

    a What do you expect to observe in this reaction?

    b What are the products of this reaction?

    c Write a word equation for this reaction.

148 What is the chemical test for hydrogen gas and what is the positive outcome which shows the gas is hydrogen?

149 What is the chemical test for carbon dioxide gas and what is the positive outcome which shows the gas is carbon dioxide?

C5 Energy changes

C5.1 Changes of state

1. Draw a particle diagram for a substance that is a solid.
2. Draw a particle diagram for that same substance after it has been melted.
3. Draw a particle diagram for that same substance after it has been boiled.
4. This question is about melting.
   a. What happens to a substance when it melts?
   b. What happens to the particles in a substance when it melts?
   c. What happens to the space between particles as the temperature increases?
   d. Explain your answer to part c.
   e. What happens to the temperature of a substance as it melts?
   f. What happens to the energy in the substance as it melts?
   g. A substance continues to be heated. What happens to the temperature of a substance after it finishes melting?
5. How can melting point data be used to identify pure substances?
6. Aluminium has a melting point of 660°C and a boiling point of 2470°C.
   a. At what temperature does aluminium melt?
   b. At what temperature does aluminium boil?
   c. At what temperature does aluminium freeze?
   d. At what temperature does aluminium condense?
   e. For each of parts a to d, explain how you know.
   f. Sketch a heating curve for aluminium.
   g. A student has a sample of pure aluminium. How could they verify that it is pure?
7. A substance has randomly arranged particles that move quickly and bounce off each other and the sides of the container.
   a. What state is the substance?
   b. How can we increase the pressure of the substance?
   c. What hazards are associated with increasing the pressure of gases?

> **Tip**
>
> For help, see Unit C3 of your Knowledge Book.

8 Use the information in the table below to answer the following questions.

Material	Melting point (°C)	Boiling point (°C)
Niobium	2469	4927
Radon	−71	−62
Rubidium	39	688
Gold	1064	2700

  a What state will niobium be at room temperature?

  b What state will gold be at 1000°C?

  c What state will niobium be at 2470°C?

  d What state will radon be at −50°C?

  e What state will radon be at −270°C?

  f What state will gold be when niobium melts?

  g At what temperature will radon melt?

  h 15 g of gold is placed into a furnace. The gold is melted so it can be poured into a mould.

    i What is the minimum temperature of the furnace?

    ii What is the maximum temperature of the furnace?

    iii What mass of molten (melted) gold will be formed?

    iv Explain your answer to part iii.

  i The record maximum temperature for England is 40°C.

    i What state would rubidium be at this temperature?

    ii What properties would we expect from rubidium at this temperature?

  j Put the materials in order from lowest boiling point to highest boiling point.

9 Carbon dioxide becomes a gas at −78.5°C, while oxygen has a boiling point of −183°C. The boiling point of carbon is so high, scientists have only been able to predict it. They predict it will boil at approximately 4800°C.

  a Draw a bubble diagram for carbon dioxide.

  b Write a word equation for the reaction between carbon and oxygen to form carbon dioxide.

  c Write a symbol equation for the same reaction.

  d Explain why carbon dioxide has a different melting point to its constituent elements.

  e Carbon dioxide can also be formed by reacting a fuel with oxygen.

    i Give an example of a fuel.

    ii When a fuel reacts with oxygen, what name is given to this type of reaction?

    iii What other factor is required for this reaction to proceed?

    iv Write a word equation and a balanced symbol equation for the formation of carbon dioxide from methane (CH_4) and oxygen.

  f Where in an atom of oxygen will the protons be found?

  g How many electrons does an atom of oxygen have?

  h Draw the electronic configuration of oxygen.

  i What was Dalton's theory of the structure of the atom?

10 Which state of matter has the greatest energy?

11 Which state of matter has the least energy?

12 A solid melts. Describe the changes in energy and motion of the particles as it melts.

13 A gas turns into a liquid.

  a What is this process called?

  b Describe the change in energy during this process.

  c Describe what happens to the temperature during this process.

  d Explain what happens to the temperature during this process.

14 This question is about reactions.

  a Give a general equation for a neutralisation reaction.

  b Identify the products formed when sodium hydroxide reacts with hydrochloric acid.

  c What is thermal decomposition?

  d Predict the products of a displacement reaction between chlorine and potassium iodide. Give a reason for your answer.

  e Why do displacement reactions occur?

> **Tip**
>
> For help, see Unit C4 of your Knowledge Book.

15 This question is about evaporation.

  a Define *evaporation*.

  b What factors can increase the rate of evaporation?

  c Suggest whether evaporation requires energy. Give a reason for your suggestion.

16 What is condensation?

17 Give an example of a substance that undergoes sublimation.

18 Describe how the arrangement of particles in a liquid differs from that in a gas.

19 This question is about pure substances.

  a Define *pure substance*.

  b How can we tell if a substance is impure?

  c How can we separate a solution?

  d Explain how filtration can be used to separate substances.

20 This question is about atoms.

  a What is an element?

  b What is an atom?

  c A student defines a molecule as 'two or more elements covalently bonded together'.

    i Explain whether they are correct.

    ii If they are not correct, write a correct version of their statement.

  d Name the three subatomic particles.

  e What is the relationship between protons and electrons in an atom?

  f What is the law of conservation of mass?

  g A student conducts an experiment, and the mass of the products is less than the mass of the reactants. Explain what has happened.

  h A student conducts an experiment, and the mass of the products is greater than the mass of the reactants. Explain what has happened.

C5.2 Endothermic and exothermic reactions

What are exothermic and endothermic reactions?

21 Which changes of state are endothermic?

22 Which changes of state are exothermic?

23 This question is about endothermic and exothermic reactions.

  a For each of the reactions in the table below, calculate the temperature change.

	Temperature before (°C)	Temperature after (°C)	Temperature change (°C)
i	21	18	
ii	1119	1019	
iii	678	645	
iv	-9	-2	
v	36	28	
vi	29	25	

    b For each reaction, identify if it is endothermic or exothermic. Give a reason for your answer.

24 A student says that only chemical reactions can be endothermic or exothermic. Explain whether they are right.

25 Give two examples of endothermic reactions.

26 During an exothermic reaction, describe what happens to the temperature of the surroundings.

27 An instant icepack uses an example of an endothermic reaction. Explain why it feels cold when it is activated.

28 With reference to energy, describe the difference between endothermic and exothermic reactions.

How do we calculate the mean and range?

29 A student wanted to investigate the effect of concentration on temperature change. They added 5 g of copper oxide to 100 cm³ of sulfuric acid and recorded the temperature change in the table below.

Concentration (mol/dm³)	Temperature change (°C)			Mean (°C)
	Trial 1	Trial 2	Trial 3	
0.1	5	6	5	
0.5	8	8	7	
1.0	12	11	9	
1.5	15	16	15	

    a Write a line of inquiry question for this investigation.

    b Give a word equation for the reaction between copper oxide and sulfuric acid, forming copper sulfate and water.

    c Write a symbol equation for this reaction.

    d Is a pure substance formed after this reaction? Explain your answer.

    e How many atoms are present in copper oxide?

f How many elements are present in copper oxide?

g Copper oxide is basic. What type of reaction is this?

h Copper oxide is a black powder. Copper sulfate is bright blue. Give two ways the student can tell a chemical reaction has occurred.

i Copy and complete the table to calculate the mean temperature change for each concentration.

j Which concentration has the largest temperature range?

k Explain whether this reaction is endothermic or exothermic.

l The student gives the symbol equation as

$$CuO(s) + H2So_4(aq) \rightarrow CuSo_4(l) + H^2O(l)$$

List as many errors as you can with this equation.

m The student wants to display their results in a graph. Suggest whether a bar chart or a line graph would be more appropriate. Give reasons for your answer.

n Why is it important that the student keeps the mass of copper oxide the same throughout the experiment?

o Why is it important that the student keeps the volume of the acid the same throughout the experiment?

p A second student recorded a set of results for the same experiment, as shown below.

Concentration (mol dm³)	Temperature change (°C)
0.5	10
1.0	13
1.5	17
2.0	20

What type of error does the student have in their data? Give a reason for your answer.

> **Tip**
> For help, see Unit C4 of your Knowledge Book.

30 A teacher reacts hydrated cobalt chloride ($CoCl_2 \cdot 6H_2O$) with thionyl chloride ($SOCl_2$) to form anhydrous cobalt chloride, hydrochloric acid and sulfur dioxide.

This is the reaction:

$$CoCl_2 \cdot 6H_2O(s) + 6SOCl_2(l) \rightarrow CoCl_2(s) + 12HCl(g) + 6SO_2(g)$$

a For each reactant and product, give the state.

b Draw a particle diagram for $CoCl_2$.

c Compare the energy of the particles in $SoCl_2(l)$ to the energy of the particles in $HCl(g)$.

d Compare the motion of the particles in $CoCl_2(s)$ to the motion of the particles in $SO_2(g)$.

e What precautions should be taken when conducting this experiment?

f Why is this experiment not commonly done in school science laboratories?

g This reaction is endothermic. Suggest what the teacher might observe during this reaction.

31 These questions are about pressure.

  a Describe the relationship between temperature and pressure.

  b Using the following key words, explain why it is dangerous to throw a can of deodorant into a fire.

  Key words: *combust* or *combustion*; *energy*; *exothermic* or *endothermic*; *pressure*; *collision*.

32 A student wanted to investigate which metals cause the greatest temperature change when they react with hydrochloric acid.

  a Write a line of inquiry question for this investigation.

  b State the independent variable.

  c State the dependent variable.

  d What factors should be kept the same during this investigation?

  e Why is it important to keep these factors the same?

  f The student says they need to 'keep the amount of acid the same'.

   i What is wrong with this statement?

   ii Write a correct version of this statement.

  g The student's results are shown below.

Metal	Temperature change (°C)		
	Trial 1	Trial 2	Mean
Copper	0	0	
Zinc	12	12	
Iron	10	7	
Magnesium	15	14	
Lead	7	8	

   i Copy and complete the table to calculate the mean temperature change for each metal.

   ii Which metal caused the greatest temperature change?

   iii Which result is anomalous? Use data from the table to support your answer.

iv Which metal has the greatest temperature change?

v Which metal did not react with the acid? Use data from the table to support your answer.

vi The bigger the temperature change, the more reactive the metal is. Use data from the table to put the metals in order from most reactive to least reactive.

h A different student says they should have used a polystyrene cup instead of a glass beaker to conduct this investigation. Suggest why.

i Write a word equation for each of the reactions in this investigation.

j The student draws a line graph to represent their results. Explain why they are wrong.

k When metals react with acid, explain whether it is an endothermic or an exothermic reaction.

33 Use the table below to answer the following questions.

Mass of ammonium chloride added (g)	Temperature change (°C)		
	Trial 1	Trial 2	Mean
10	−7.0	−6.0	
20	−13.2	−12.8	
30	−19.0	−17.0	
35	−20.0	−5.3	
40	−20.1	−20.0	

a Copy and complete the table to calculate the mean for each mass.

b Which result is anomalous?

c Draw a graph of the mean temperature change against the mass of ammonium chloride added.

d Why does the reaction temperature begin to increase?

e Ammonium chloride has the following hazard symbol.

What precautions should be taken when working with ammonium chloride?

f A teacher adds some universal indicator to the product. It turns yellow-orange. What pH is the product?

34 A teacher heats green copper carbonate in a test tube until it turns into black copper oxide and carbon dioxide.

   a How can you tell a chemical reaction has taken place?

   b 8 g of copper carbonate is heated. Would you expect the mass in the test tube to increase, decrease or stay the same?

   c Explain your answer to part b.

   d This reaction is a thermal decomposition reaction. Is a thermal decomposition reaction an *exothermic* or an *endothermic* reaction?

   e Other than the colour change, what observations could be possible during this reaction?

   f Write a word equation for this reaction.

   g Write a symbol equation for this reaction. Include state symbols in your answer.

P4 Pressure in fluids

P4.1 Pressure in liquids

How does the pressure felt change as we go deeper into water?

1. Which state of matter best fits the description of its particles below? Some might have more than one answer.
   a. The particles are regularly arranged.
   b. The particles are randomly arranged.
   c. The particles can only vibrate.
   d. The particles are touching.
   e. There are spaces between the particles.
   f. The particles are free to move.

2. What do we call substances that can flow?

3. Which states of matter can flow?

4. When a cup of water is spilled, the puddle is a different shape to the cup. Why does this prove that water is a fluid?

5. What causes a liquid to exert a pressure on a container?

6. What causes a gas to exert a pressure on a container?

7. A student gets two 1L plastic bottles. They fill one bottle with air by opening and closing the lid. They fill the other bottle with water. Suggest which one has the greatest pressure exerted on the inside of the bottle. Give a reason for your answer.

8. What happens to the pressure as you go deeper into water?

9. What causes the sea to exert pressure on a submarine?

10. What causes the pressure on your ears to increase if you go down to the bottom of the deep end of the swimming pool?

11. A dam has been built to create a reservoir.
    a. How will the water pressure change as it gets deeper? Give a reason for your answer.
    b. Explain why the dam is wider at the bottom than at the top.

12. A student says, 'Sand must be a liquid because the sand grains can flow.' Explain why they are wrong.

> **Tip**
>
> These questions also cover 'What are fluids?' and 'What causes the pressure of a liquid in a container?'

13 There are two submarines. Submarine A is near the surface. Submarine B is deep down by the ocean floor.

   a Which one will experience the most pressure?

   b Why are the walls in a submarine thicker than those in a boat?

Why do objects float?

14 What is the name of the contact force that is created by the pressure of a liquid on a boat?

15 When an object floats, what do we know about the size of the weight and upthrust?

16 Is weight a contact or non-contact force? Give a reason for your answer.

17 What are the two states of matter that are fluids?

18 What physical property do fluids have?

19 You swim to the bottom of the deep end of the swimming pool. Explain why the pressure increases. Include the term *particles* in your answer.

20 A student is experimenting with some polymer clay and a tank of water.

   a Some clay is submerged in a tank of water. The clay sinks. What do we know about the size of the weight and upthrust?

   b The clay takes 3.3 seconds to hit the bottom of the tank from when it was released. The water has a depth of 66 cm. Calculate the speed of the clay in metres per second. Use the EVERY method to show your working.

   c The clay is now spread out into a boat shape. The clay floats. Explain how changing the shape of the clay caused it to float.

   d The clay is made into a star shape. It has a weight of 5 N. The upthrust on the clay is 4.5 N.

      i Calculate the resultant force.

      ii Describe how the motion of the clay changes.

21 An empty can is filled with air and sealed with a lid. The can is tied to a weight and thrown into the sea.

   a Describe the particle arrangement inside the can.

   b Describe the particles' motion inside the can.

   c The air inside the can creates a pressure. Describe how the air particles create the pressure inside the can.

   d Describe how the water on the outside exerts a pressure on the sides of the can.

> **Tip**
>
> For help with forces, see Unit P3 of your Knowledge Book.

> **Tip**
>
> 1 m = 100 cm

e As the can sinks, what can we say about the size of the upthrust and weight?

f As the can reaches the bottom of the ocean its sides begin to crumple. There is air in the can pushing outwards and water outside pushing inwards. Explain why the can gets crushed as it gets deeper in the ocean.

g What happens to the gravitational energy store as the can sinks?

P4.2 Atmospheric pressure

What is atmospheric pressure?

22 The diagram below shows a box with gas particles inside it.

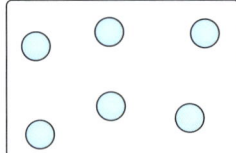

a Describe the motion of the particles in the box.

b Draw a box that would have a higher density of particles.

c Draw a box that would have particles with a lower pressure.

d What causes the pressure on the sides of the container?

23 What is the atmosphere?

24 A student says, 'When the atmospheric pressure is increased it is because the particles are colliding more with my body than they are with each other.' Are they correct? Give a reason.

Why does atmospheric pressure change?

25 What do we call the layer of gases that surrounds the Earth?

26 Describe the arrangement of particles in a gas.

27 Describe the motion of particles in a gas.

28 Explain how gas particles can exert a pressure on an object.

29 A student says, 'As a person goes up a hill the atmospheric pressure will drop because there is less weight of the air pushing them down.' Are they correct? Give a reason for your answer.

30 How does increasing an object's height above the ground affect the atmospheric pressure it feels?

Tip

These questions also cover 'How does atmospheric pressure change?'

31 A mountaineer is attempting to reach the top of Mount Everest.

   a What will happen to the number of air particles in a fixed volume as they climb?

   b What will happen to the number of collisions between the air particles and the mountaineer as they climb?

   c A student says, 'As they climb, the density of air particles will decrease because the particles stop moving.' Are they correct? Give a reason for your answer.

   d What is the relationship between density of the air and the weight of the atmosphere on an object?

   e Name the energy store of the climber that is decreased as they climb higher.

   f Name the energy store of the climber that is increased as they climb higher.

   g Name the way the energy is transferred between the energy stores.

   h The mountaineer's muscles have a power of 25 kW. How much energy is transferred if they climb for an hour?

   i Why will they need to bring an oxygen mask if they are going to reach the top of Mount Everest safely?

   j The mountaineer manages to climb 20.5 km on the path to the top of Everest in exactly 7 days. Calculate their average speed in metres per second. Round your answer to three decimal places.

> **Tip**
>
> For help with conversions involving time and distance, see page 19 of the 'Measurement' topic in 'Working scientifically' in your Knowledge Book.

32 Explain how and why atmospheric pressure changes with height above the ground.

33 What happens to the density of the air as we move higher up in the atmosphere? Why?

34 Place the following places in order of increasing density of particles.

   A A diver at the bottom of the sea

   B A swimmer on the surface of the sea

   C A sailor standing on the deck of a boat

   D A mountaineer on top of a mountain

   E A person on the top floor of an office block

P4 Pressure in fluids

P4

P4.3 Pressure calculations

How can we change pressure?

35 Is weight a contact or non-contact force?

36 A person has a weight of 600 N.
   a How much force is going to the ground through each foot?
   b If the person stands on one foot, will their pressure increase or decrease? Explain your answer.
   c The 600 N person now sits cross-legged on a four-legged chair. How much force will go through each leg of the chair?

37 An insect has six feet and a weight of 3 N. How much force goes through each foot?

38 A polar bear has a weight of 5000 N and is walking through the snow.
   a How much force will go through each foot if it walks on just its back legs?
   b How much force will go through each foot if it walks on all four legs?
   c Explain why the polar bear finds it easier to walk on the snow on all four feet rather than just on two feet.

39 Camels have feet with large surface areas. This helps them to walk on sand. Explain why.

40 Cars that drive on snow have big, wide tyres. Explain why.

41 Nails have a very small surface area at their pointy end. Explain why.

42 If a person stands on one nail, it will go through their foot. However, if they lie down on a bed of 100 nails, they are fine. Explain why.

43 What is a fluid? Give two examples.

44 Describe the arrangement and motion of particles in a gas.

45 How do particles of a gas create pressure on the sides of a container?

46 What is the relationship between the speed of gas particles and the pressure they exert on the container?

47 Explain why a helium balloon rises in the air. Use the terms *weight* and *upthrust* in your answer.

How do we calculate pressure?

48 Write the word equation that links pressure, area and force.

49 What is the unit of pressure?

50 What is the relationship between the pressure exerted on an area and the force experienced?

51 What states of matter are fluids?

52 What is a fluid?

53 Which states of matter have a fixed shape?

54 Calculate the pressure that a force of 50 N exerts on an area of 2 m^2.

55 Calculate the pressure that an object with an area of 0.5 m^2 experiences when a force of 1000 N is applied.

56 What is the relationship between the height above the ground and the atmospheric pressure?

57 What is the relationship between the depth below sea level and the pressure the water exerts?

58 A surface has a force of 90 N acting on it and has an area of 15 m^2. What is the pressure on the surface?

59 A surface has a force of 225 N acting on it and has an area of 0.015 m^2. What is the pressure on the surface?

60 A surface has an area of 40 m^2 and a total force acting on it of 8800 N. What is the pressure on the surface?

61 At the top of a hill, the air exerts a force of 4050 N on a person who has a surface area of 1.8 m^2. Calculate the atmospheric pressure.

62 A person has a weight of 630 N and is standing on one foot. Their foot has a surface area of 0.009 m^2.

   a What is the person's pressure?

   b What is their pressure if they stand on two feet?

   c Compare your answers to parts a and b. What do you notice? Explain why this is so.

The questions below use the three equations you have learned so far for pressure, speed and power. For each question, first work out which equation you need and then answer the question.

63 A scooter is rolling down a hill.

   a It travels 6.8 m in 8 s. Calculate the speed of the scooter in metres per second.

   b A force of 750 N is pushing down on the scooter's wheels. The wheels have a total area of 0.0025 m^2. Calculate the pressure on the wheels.

> **Tip**
>
> For help with the equations involving power and speed, see Units P1 and P2 of your Knowledge Book.

P4 Pressure in fluids

c The rider pushes the scooter back up the hill. Their legs have a power of 500 W. It takes them 3 minutes to get to the top of the hill. Calculate the total energy transferred by the rider's muscles.

d What energy store in the rider is decreased while they push the scooter up the hill?

e What energy store of the rider is increased while they push the scooter up the hill?

64 A drawing pin has an applied force of 25 N on it. The pin head has an area of 0.000 001 m². Calculate the pressure.

65 A microwave has a power of 600 W and is dropped out of a second-story window 12 m above the ground. It takes 1.4 s to hit the ground. Calculate the speed of the microwave. Give your answer to two decimal places.

66 A student is attending archery practice. They fire an arrow at the target. The arrow travels 70 m to the target in 2.1 seconds. The arrow hits the target with a force of 65 N. The arrowhead has a surface area of 0.000 02 m².

a Calculate the speed of the arrow.

b Calculate the pressure of the arrow when it hits the target.

67 A submarine is under the Arctic Circle. It has a surface area of 650 m² and is powered by an engine with a power of 550 kW.

a The water exerts a total force of 9 100 000 N on the submarine. Calculate the pressure on the submarine hull.

b Calculate the speed if the submarine can travel 30 000 m in an hour. Give your answer to two decimal places.

c How much energy will the engine transfer in the hour-long journey?

> **Tip**
>
> See the 'Measurement' topic in 'Working scientifically' in your Knowledge Book for a reminder of how to convert minutes to seconds. There are 60 minutes in an hour.

68 A plane is sitting stationary on the runway waiting to take flight.

a The plane has a weight of 550 000 N. What is the value of the normal reaction force of the runway?

b The plane starts its engines and begins moving forwards. What type of force do the engines provide?

c Is the force in part b a contact or non-contact force?

d What force will work against the motion of the plane?

e The forwards resultant force is increased. What happens to the motion of the plane?

f The plane takes off. Name the force that causes the plane to take flight.

g Is the force in part f a contact or non-contact force?

h The plane travels 15 km in 1 minute 15 seconds. Calculate the speed of the plane in metres per second.

i The plane has four engines. Each has a power of 150 kW. Calculate the total energy transferred by the engines when the plane flies for 30 seconds.

> **Tip**
>
> For help with unit conversions involving the prefix 'kilo', see Topic P1.1 in your Knowledge Book.

69 Use the idea of particles to explain why a mountaineer might suffer from dizziness as they climb to higher altitudes.

70 Why does a ship made of steel, which is denser than water, float?

71 What is upthrust? How does it relate to an object's ability to float or sink?

72 Why do your ears 'pop' when you ascend or descend rapidly in an plane?

P5 Sound

P5.1 Types of wave

What do waves do?

1. Give two examples of a wave.
2. What do waves transfer from one place to another?
3. What are the standard units of energy?
4. What do waves not transfer from one place to another?
5. What is a medium?
6. Name the four transfers of energy.
7. Give a way in which we know that waves do not transfer particles.
8. Which medium can sound not travel through?
9. A student has drawn the graph below but made two mistakes. Explain each mistake and how to fix it.

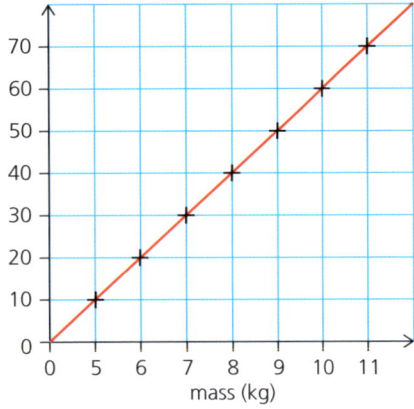

What are waves made of?

10. Give an example of an object that oscillates.
11. What is an oscillation?
12. A student says, 'Water waves transport particles because particles of water move.' Explain why they are wrong.
13. What do sound waves, light waves and water waves have in common?
14. Which type of wave can travel in a vacuum: light waves or sound waves?

15 A speaker has a power of 50 W.

   a How many joules of energy does it transfer per second?

   b The speaker is left on for 2 minutes. How much energy is transferred in this time?

   c What happens to the particles in the air as the sound travels through the air?

   d The sound wave comes to a wall. Can the sound travel through the wall? Explain your answer.

> **Tip**
>
> For help, see the Worked example in Topic P1.3 of your Knowledge Book.

16 A student attaches a mass to a spring and lets it go.

   a What force pulls the mass towards Earth?

   b What happens to the elastic force as the spring stretches more?

   c When the mass is let go, how would we describe the motion of the mass?

   d The spring constant of the spring is 250 N/m. The spring is stretched at one point by 0.2 m. What is the force on the spring?

> **Tip**
>
> For help, see the Worked example in Topic P3.4 of your Knowledge Book.

What types of wave are there?

17 What are the names of the two types of wave?

18 How are the direction of the oscillation and the direction of the wave related in a transverse wave?

19 What is oscillating in a sound wave when:

   a it travels through the air

   b it travels through a wall

   c it travels through water?

20 How are the direction of the oscillation and the direction of the wave related in a longitudinal wave?

21 Give three examples of a transverse wave.

22 What do transverse and longitudinal waves have in common?

23 What is the main example of a longitudinal wave?

24 Draw a transverse wave and label the peak and trough.

25 In a longitudinal wave, what is a compression?

26 In a longitudinal wave, what is a rarefaction?

27 Sound wave 1 travels at 330 m/s in air. Sound wave 2 travels at 1100 m/s in water. What is the relative speed of sound wave 1 to sound wave 2 if:

   a they travel in the same direction

   b they travel in opposite directions?

> **Tip**
>
> For help, see the Worked example in Topic P2.3 of your Knowledge Book.

28 Here is a line of inquiry: How does the age of a runner affect how fast they are?

   a What is the independent variable?

   b What is the dependent variable?

How do we measure the features of a transverse wave?

29 Measure the amplitude and wavelength of each of the waves below.

a
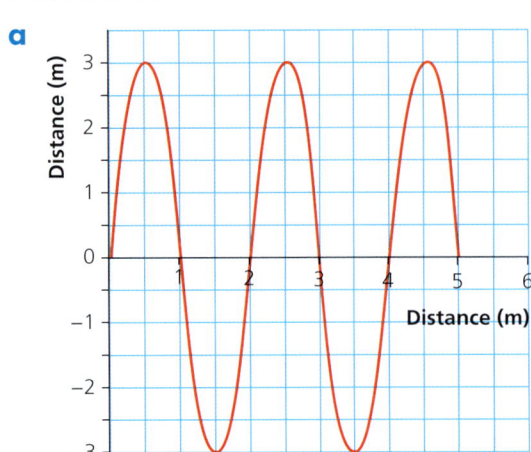

c
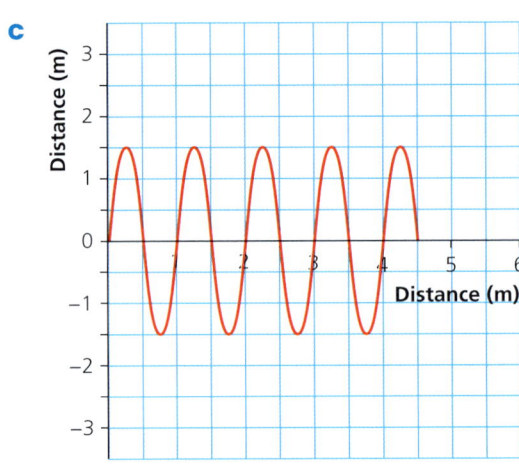

b
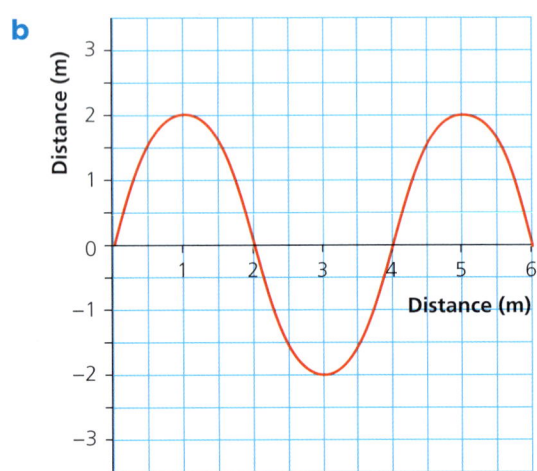

d
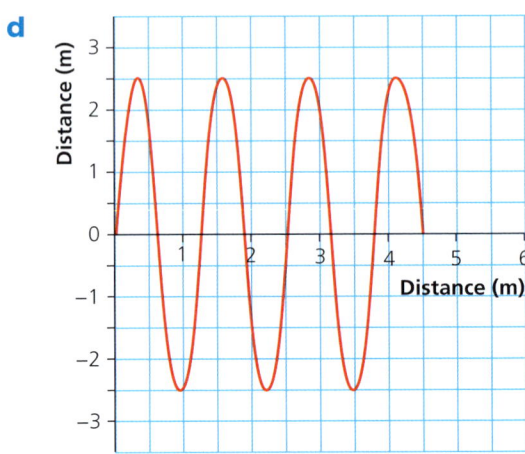

30 How do you find the amplitude of a wave?

31 How do you find the wavelength of a wave?

32 Below is a diagram of a wave.

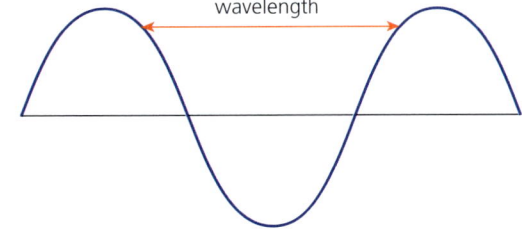

Tip

These questions also cover 'What are the features of a transverse wave?'

    a What type of wave is shown?

    b What is wrong with the way the wave has been labelled?

33 What happens to light if we increase the amplitude?

34 Large stones are dropped into a pond, followed by smaller and smaller stones. What happens to the amplitude of the water waves?

35 A student says, 'To measure the amplitude, we go from the trough of a wave to the peak.' Explain why they are wrong.

36 How can we prove that a water wave does not transport particles from one place to another?

37 Explain what happens to the pressure in a liquid as we go deeper into a body of water.

38 What is the definition of a *medium* when talking about waves?

How can waves behave?

39 Describe how reflection occurs.

40 What is a reflection of sound called?

41 A student labels the diagram below. What is wrong with what they have done?

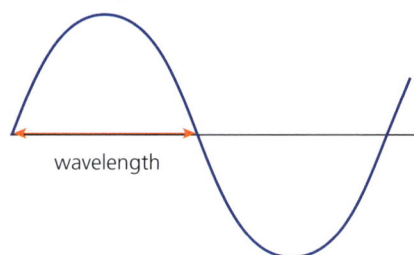

42 Measure the amplitude and wavelength of each of the waves below.

a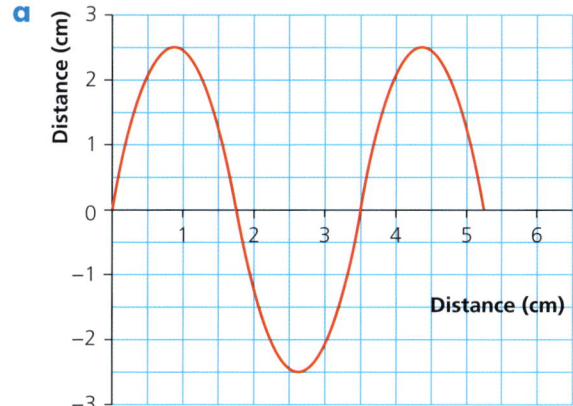

b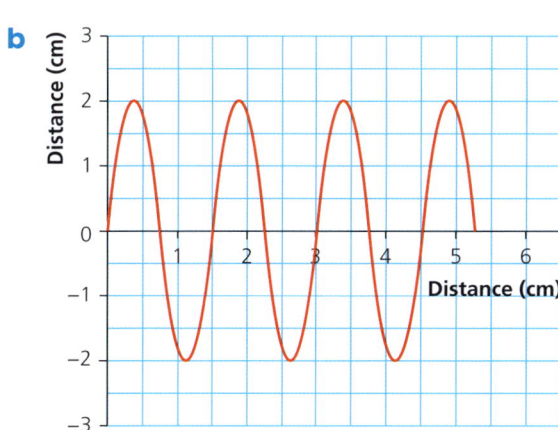

c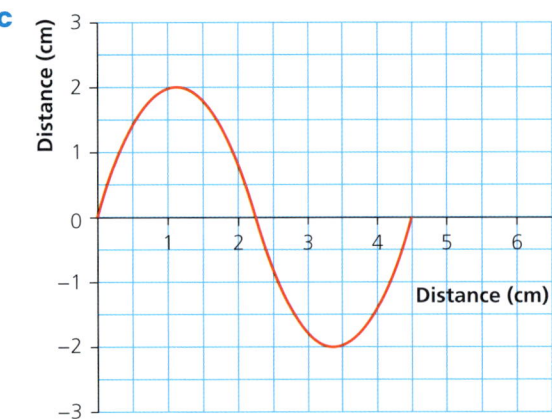

d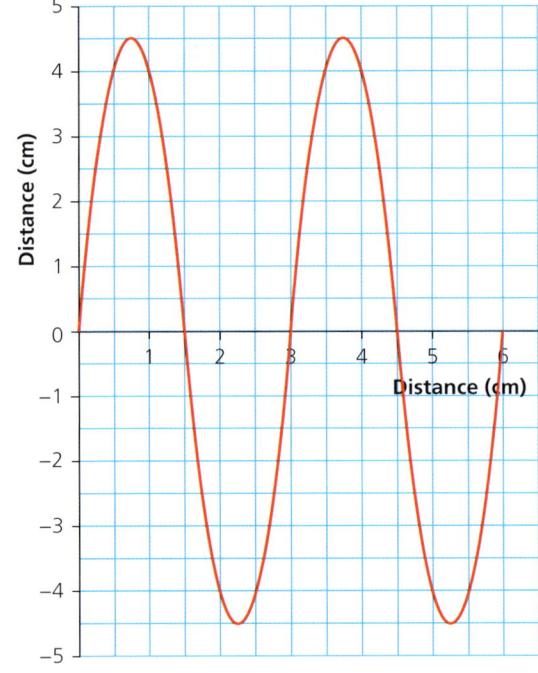

43 Not all waves are the same.

   a What are the names of the two types of wave?

   b Describe how they are different from one another.

   c Describe how they are similar to one another.

44 A student has a battery-powered torch. The torch has a power of 25 W.

   a Which store of energy is decreasing when the torch is turned on?

   b What is the name for the transfer of energy between the battery and the bulb?

   c What is the name for the transfer of energy between the bulb and the environment as it emits light?

   d What type of wave is light?

   e The torch is left turned on for 1 minute. How much energy is transferred by the torch?

> **Tip**
> Remember the units.

45 A student measures the height from a peak to a trough and says this is the amplitude. Explain why they are wrong.

46 A student wants to test whether the mass of water affects how long it takes to freeze.

   a Write a line of inquiry for this experiment.

   b What is the independent variable?

   c What is the dependent variable?

   d What are some control variables they need to consider?

P5.2 Sound waves

Where can sound waves travel?

47 What is oscillating when sound travels?

48 What kinds of medium can sound travel through?

49 What is a *vacuum*?

50 Which states of matter are considered fluids?

51 Why can sound not travel in a vacuum?

52 What is the speed of sound in air?

53 Does sound travel faster in solids, liquids or gases?

54 Why does sound travel faster in solids than in gases?

55 When in steel, it takes 1.2 s for sound to travel a distance of 6 km. How fast does sound travel in steel?

56 When in air, it takes 1.2 s for sound to travel a distance of 396 m. How fast does sound travel in air?

Tip

Remember the units.

Tip

For help, see the Worked example in Topic P5.2 of your Knowledge Book.

What is frequency?

57 What is the definition of the *frequency of a wave*?

58 What are the units of frequency?

59 A duck bobs up and down on the water.

  a If it bobs up and down two times per second, what is the frequency?

  b If it bobs up and down three times per second, what is the frequency?

  c If it bobs up and down five times in 2 s, what is the frequency?

  d If it bobs up and down 12 times in 10 s, what is the frequency?

60 What is the *wavelength of a wave*?

61 Why does sound travel slower in a gas than in a liquid?

62 What is the definition of a *longitudinal wave*?

63 Name the five main stores of energy.

64 In a sound wave, what is:

  a a compression

  b a rarefaction?

65 When in water, sound can travel a distance of 3650 m in 7.3 s. How fast does sound travel in water?

Tip

For help, see Topic P1.2 of your Knowledge Book.

P5

How can we change a sound wave?

66 What is the relationship between the amplitude of a sound wave and its volume?

67 What are the units that we use for frequency?

68 What is the relationship between the frequency of a wave and the pitch of a sound?

69 Four different sound waves travel through the air.

  a Sound A: The air particles oscillate 400 times per second. What is the frequency?

  b Sound B: The air particles oscillate 350 times per second. What is the frequency?

  c Sound C: The air particles oscillate 40 times per second. What is the frequency?

  d Sound D: The air particles oscillate 2500 times per second. What is the frequency?

  e Order the four sounds from highest pitch to lowest pitch.

70 How can we increase the volume of a sound wave?

71 Below are traces of four sound waves.

i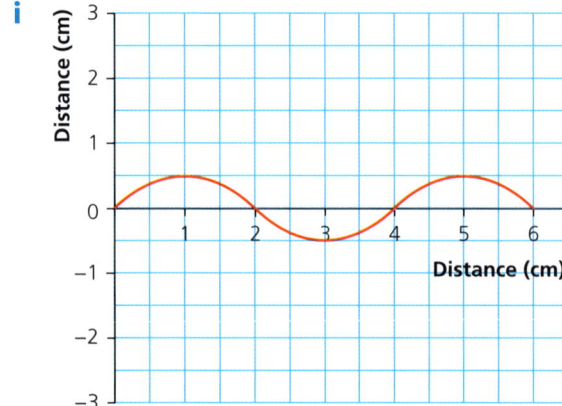

iii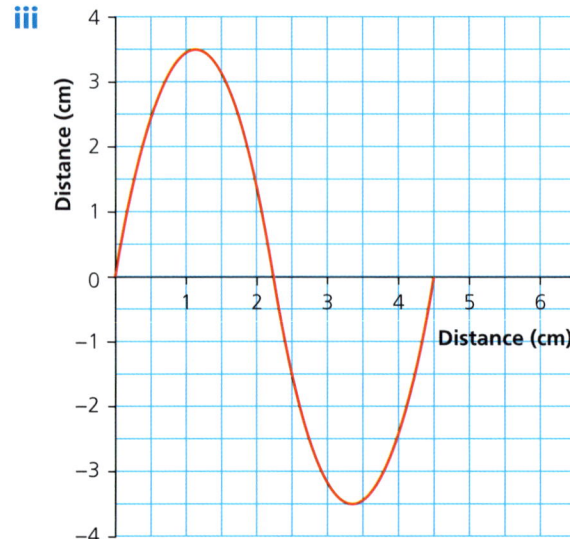

ii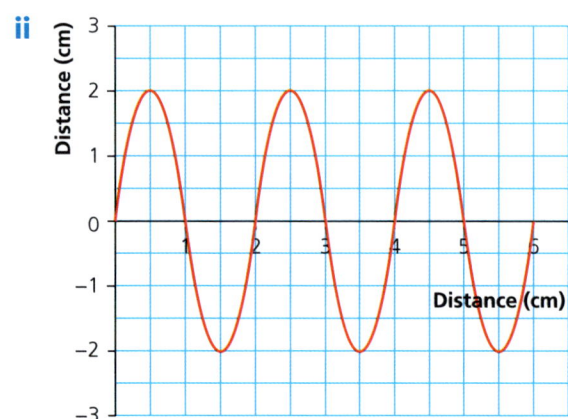

iv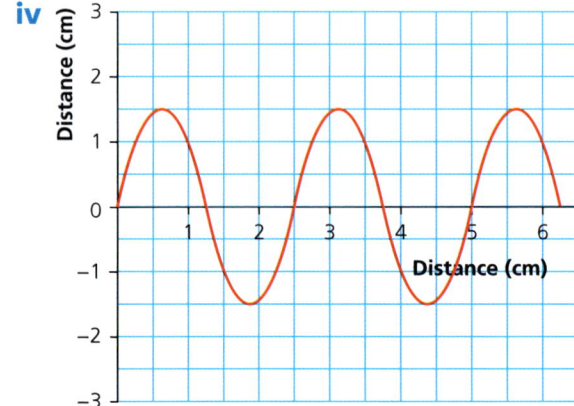

a Measure the amplitude and wavelength of each of the waves i–iv.
  b Which sound trace is for a sound with the highest volume?
  c Which sound trace is for a sound with the lowest pitch?
72 How do we make a sound lower in pitch?

How can we make sounds?

73 Explain how a drum makes a sound wave.
74 Explain how your voice makes a sound wave.
75 How are all kinds of sound produced?
76 Why do we often show sound as a transverse wave even though it is longitudinal?
77 As a sound wave passes by, the amplitude decreases. What would we notice about the sound as this happens?
78 As an ambulance passes by, the pitch of the sound wave decreases and the note gets lower. What does this mean about the frequency of the oscillations of particles in the air?
79 Four different sound waves travel through the air.
  a Sound A: The air particles oscillate 3500 times per second. What is the frequency?
  b Sound B: The air particles oscillate 8000 times in two seconds. What is the frequency?
  c Sound C: The air particles oscillate 9000 times in three seconds. What is the frequency?
  d Sound D: The air particles oscillate 10 000 times in four seconds. What is the frequency?
  e Order the four sounds from lowest pitch to highest pitch.
80 Explain how an engine makes a sound wave.

How does our ear work?

81 Which part of the ear vibrates to detect sound?
82 Put statements A–F in the correct order to explain how our ears detect sound.
  A The vibrations of the ear drum are passed to the cochlea.
  B The particles of air hit the ear drum.
  C A sound wave in the air travels down the auditory canal through the air in there.
  D The electrical signal is passed down the auditory nerve to the brain.

> **Tip**
> These questions also cover 'How can we hear sounds?'

E The ear drum vibrates.

F The cochlea turns the vibrations into an electrical signal.

83 Name each part of the ear in the diagram below.

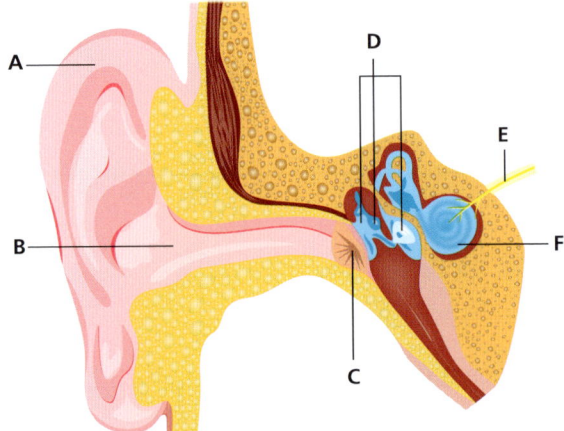

84 What would happen if we did not have an ear drum?

85 Give another word for *vibrate*.

86 a Name the two types of wave.

b Describe the difference between them.

87 Waves transfer energy.

a What are the names of the four transfers of energy?

b Which transfer of energy is happening between a sound wave and the ear drum?

c Which transfer of energy occurs between the cochlea and the brain?

d What do waves *not* transfer?

e A sound wave has a power of 0.02 W and I hear the sound for 3 s. How much energy does the sound wave transfer to me?

88 If we put some rice on a drum skin and hit the drum, the rice bounces up into the air.

a What is a force?

b What is the name of the force that pulls the grain of rice back down to Earth?

c What units do we use to measure force?

d As the grain of rice moves through the air, what other force might it experience?

89 Sound travels at 5000 m/s in steel (a solid), but at only 330 m/s in air (a gas).

a Why does sound travel faster in steel than in the air?

b How far will sound travel in steel in 1 s?

c How far will sound travel in air in 1 s?

d What is the relative speed of sound in steel to sound in air if the waves are travelling in the same direction?

   e What is the relative speed of sound in steel to sound in air if the waves are travelling in opposite directions?

90 A child jumps up and down on a trampoline.

   a What word describes their motion as they jump up, and down, and up, and down?

   b Draw a force diagram of the forces acting on the child before they start jumping, when they are just standing on a trampoline.

   c Draw a force diagram of the forces acting on the child when they are in the air.

   d What three things can forces do to moving objects?

How do we identify trends in a table?

91 Which variable goes in the first column of a table: the independent variable or the dependent variable?

92 What information must go in the headings of the columns of a table?

93 Which variable will we know the numbers for before we start the investigation and why?

94 A student wants to do an investigation into how bright a bulb is when they add more batteries. They write the following line of inquiry: 'How does the brightness change?'

   a This is not a proper line of inquiry. Correct their line of inquiry.

   b Write a prediction for their line of inquiry.

   c What type of wave is light?

   d What transfer of energy is happening between the battery and the bulb?

   e What transfer of energy is happening between the bulb and the rest of the room?

> **Tip**
>
> See 'Lines of inquiry and variables' on page 7 of your Knowledge Book.

95 Look at the table below.

Density of medium (kg/m³)	Speed of sound (m/s)
1.2 (Air)	343
997 (Water)	1480
2000 (Brick)	4200
2500 (Glass)	4540
8830 (Copper)	4600

a Write the line of inquiry for this investigation.

b What is the trend from the table?

c Explain why this might be, looking at the names of the materials to help.

96 Look at the table below.

Age of Stacey (years)	Height of Stacey (m)
12	1.2
13	1.3
14	1.4
15	1.5
16	1.6
17	1.6
18	1.6

a Write the line of inquiry for this investigation.

b What is the trend from the table?

97 Look at the table below.

Power of kettle (W)	Time it takes water to boil (s)
1000	350
1200	300
1400	250
1600	200
1800	150
2000	100
2200	50

a Write the line of inquiry for this investigation.

b What is the trend from the table?

c Suggest a control variable in this experiment.

d Suggest why the trend is this way.

What sounds can we hear?

98 What is the frequency of the lowest note that humans can hear?

99 What is the frequency of the highest note that humans can hear?

100 Calculate the range of the values for human hearing.

101 What do we call sounds above the highest frequency that humans can hear?

102 What do we call sounds below the lowest frequency that humans can hear?

103 The amplitude of a sound wave is increased.

  a Define *amplitude*.

  b What happens to the sound wave when the amplitude increases?

  c What happens to the vibrations of the ear drum as the amplitude of the sound wave increases?

104 Name an animal that can hear ultrasound.

105 Name an animal that can hear infrasound.

106 A student wants to investigate what happens to the highest frequency that we can hear as we get older.

  a Write a line of inquiry for this investigation.

  b Write a prediction for this investigation.

  c The student takes five 73-year-olds and measures the highest frequency they can hear. The results are 16 500 Hz, 15 000 Hz, 17 000 Hz, 16 500 Hz and 14 000 Hz.

   i Calculate the mean of their results.

   ii Calculate the range of their results.

107 Look at the table below.

Lowest frequency an animal can hear (Hz)	Hearing range (Hz)
2	1500
4	2000
12	210 000
20	20 000
100	35 000
200	40 000
9000	190 000

  a Write the line of inquiry for this investigation.

  b Are there any anomalies in the table?

  c What is the trend from the table?

108 Explain what happens to the pressure from the air as you get higher up a mountain.

109 Below is a graph of the motion of a water wave out at sea.

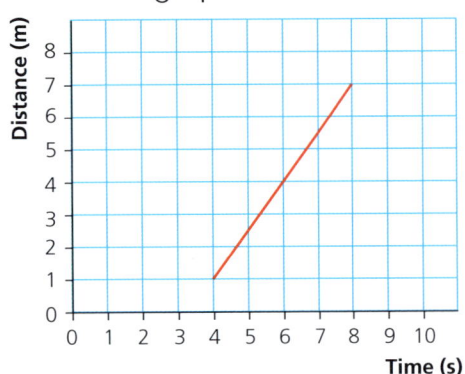

a What distance is the wave when we start measuring?

b What distance is the wave when we stop measuring?

c How far does the wave travel?

d At what time do we start measuring the wave?

e At what time do we stop measuring the wave?

f How long is the wave measured for?

g What is the speed of the wave?

How can we measure the speed of sound?

110 A speaker is 70 m from a wall. A student is standing next to the speaker when it begins to make a sound. They hear the echo 0.4 s later.

a How far does the sound travel?

b What is the speed of the sound wave?

111 A naval officer is underwater in a submarine. The submarine emits a sound. It is 2000 m from a cliff. They hear the echo 3.2 s later.

a How far does the sound wave travel?

b What is the speed of the sound wave?

112 A mountaineer is standing 250 m from a mountain. They shout and 1.6 s later they hear the echo.

a How far did the sound wave travel?

b What is the speed of sound?

113 A student has the following line of inquiry: 'How does the volume of water affect how long it takes to freeze?'

a What is the independent variable?

b What is the dependent variable?

c Write a prediction for this line of inquiry.

Tip

For help, see the Worked example in Topic P5.2 of your Knowledge Book.

114 Look at the table below.

Temperature of water (°C)	Speed of sound in water (m/s)
10	1480
15	1500
20	1520
25	1200
30	1580
35	1620
40	1640

a Write the line of inquiry for this investigation.

b Are there any anomalies in the table?

c What is the trend from the table?

P5.3 Microphones and ultrasound

How do microphones work?

115 What is a diaphragm?

116 How does a microphone work?

117 In what ways is a microphone similar to an ear?

118 How are the direction of the oscillation and direction of the wave related in a longitudinal wave?

119 What do waves *not* transfer from one place to another?

120 Sound does not travel at the same speed in every medium.

  a What is a medium?

  b In which state (solid, liquid or gas) does sound travel fastest?

  c Explain your answer to part b.

121 A diaphragm is made of elastic material.

  a What does *elastic* mean?

  b What is the name of the force when there is stretched elastic material?

  c What happens to the force from part b as the material is stretched more and more?

  d What are the units we use to measure force?

122 What will happen to a microphone if it detects a sound that is getting louder?

123 The diaphragm in a microphone starts to vibrate less frequently. What does this mean about the note that it is detecting?

P5

How do speakers work?

124 How does a speaker work?

125 What is the definition of the frequency of a wave?

126 What are the units we use to measure frequency?

127 The diaphragm in a speaker oscillates to produce a sound.

   a It oscillates 100 times a second. What is the frequency?

   b It oscillates 2000 times a second. What is the frequency?

   c It oscillates 12 000 times in four seconds. What is the frequency?

128 A speaker's diaphragm oscillates four times a second. Why can nobody hear it?

129 Explain what happens inside a speaker to produce a louder sound when we increase the volume of the music playing through it.

130 A musician plays a sound in a large stadium from a speaker at the front. They stand at the back of the stadium, 99 m away. The sound takes 0.3 s to travel to their ears.

   a Calculate how fast the sound is travelling.

   b Would the sound travel faster or slower if the stadium was entirely underwater?

   c Explain your answer to part b.

What is ultrasound?

131 What is ultrasound?

132 The diaphragm in a speaker oscillates to produce a sound.

   a It oscillates 15 000 times a second to produce sound A. What is the frequency?

   b It oscillates 23 000 times a second to produce sound B. What is the frequency?

   c It oscillates 18 000 times a second to produce sound C. What is the frequency?

   d It oscillates 50 000 times a second to produce sound D. What is the frequency?

   e Which sounds can humans hear?

133 Name three animals that can hear ultrasound.

134 How is ultrasound similar to 'normal' sound?

135 How is ultrasound different to 'normal' sound?

136 A dog whistle emits sound that most humans cannot hear, but dogs can. What does this mean about the sound?

137 What is oscillating in a sound wave when:

   a it travels through a piece of metal

   b it travels through a swimming pool

   c it travels from the speaker in my room to my ear?

What else do we use ultrasound for?

138 How can we use ultrasound to find the distance to an object?

139 A fishing boat has ultrasound equipment on it. What happens to the ultrasound waves when they hit a shoal of fish?

140 In water, ultrasound travels at 1500 m/s. It takes 1 s for the waves to travel from A to B and back.

   a How far has the ultrasound wave travelled?

   b The distance from A to B is only 750 m. Why is this?

141 Some animals use echolocation to find prey or obstacles. How does this work?

142 What is the definition of *pressure*?

143 When in a sound wave is the pressure highest: at a compression or a rarefaction?

144 Why are sound waves sometimes called pressure waves?

145 How can ultrasound waves be used to clean?

146 Why can we not hear ultrasound?

147 Measure the amplitude and wavelength of each of the waves below.

> **Tip**
>
> These questions also cover 'How do we use ultrasound?'

a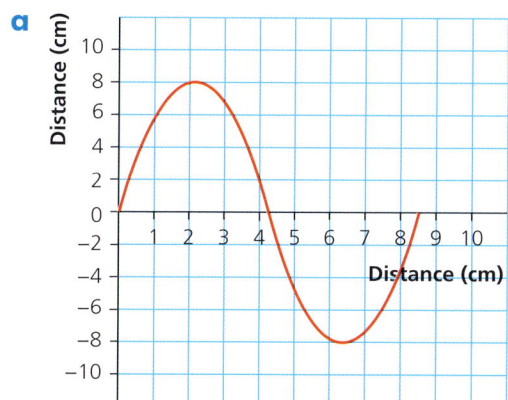

b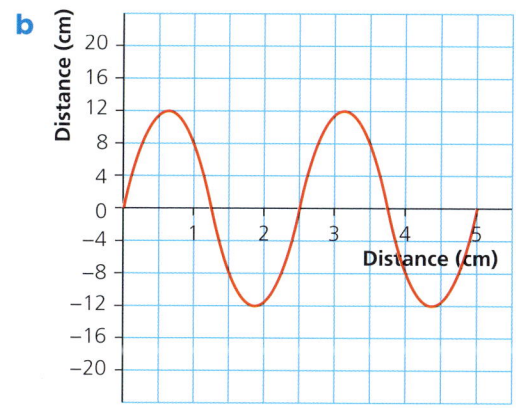

c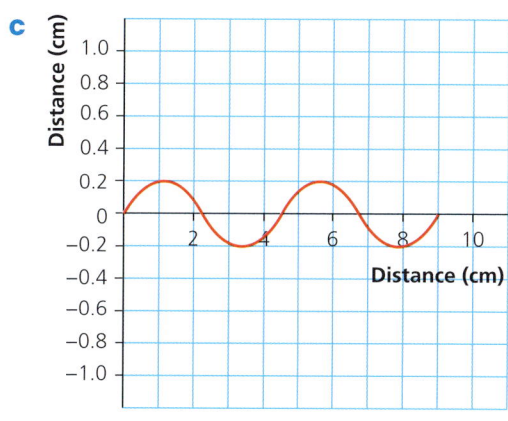

d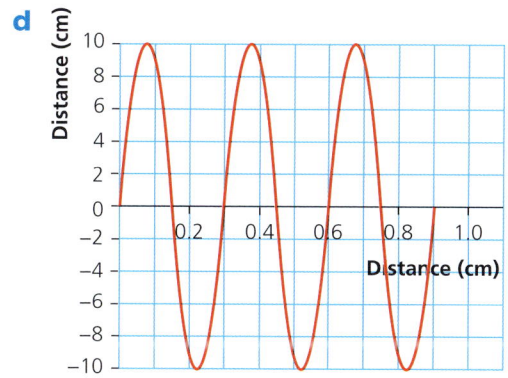

P6 Light

P6.1 Light and ray models

Where can light travel?

1. What is the name of the energy transfer when light transfers energy?
2. What is a vacuum?
3. Why can light travel through a vacuum?
4. How do we detect light?
5. Name some materials that light can travel through.
6. Measure the amplitude and wavelength of each of the waves below.

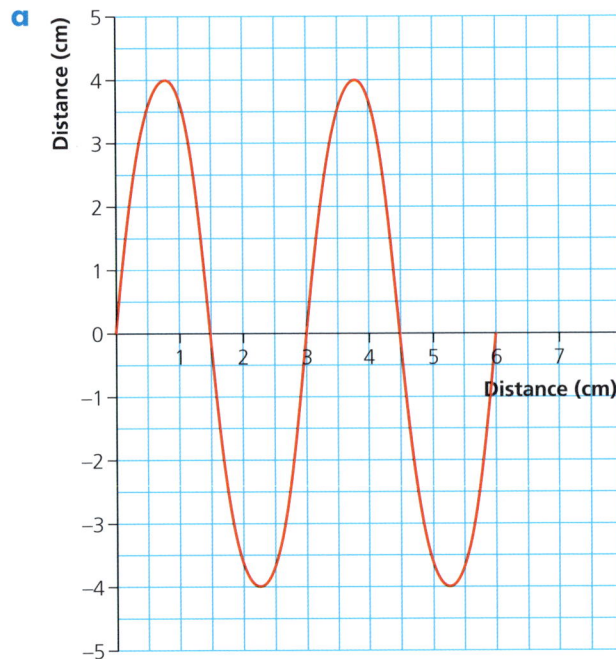

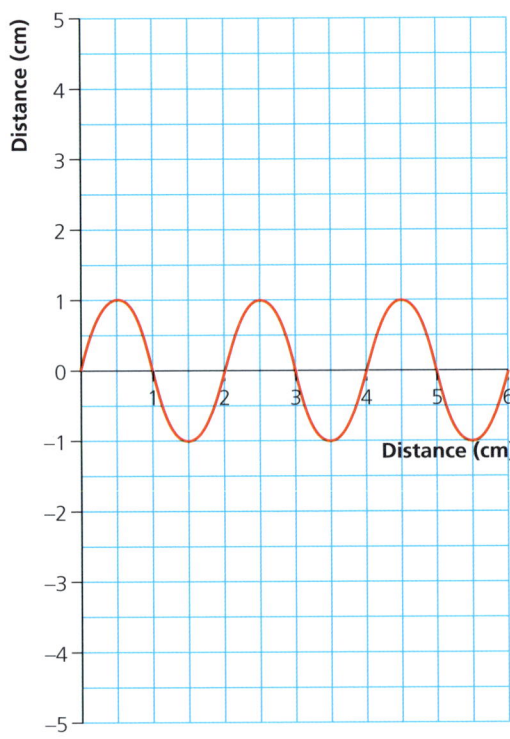

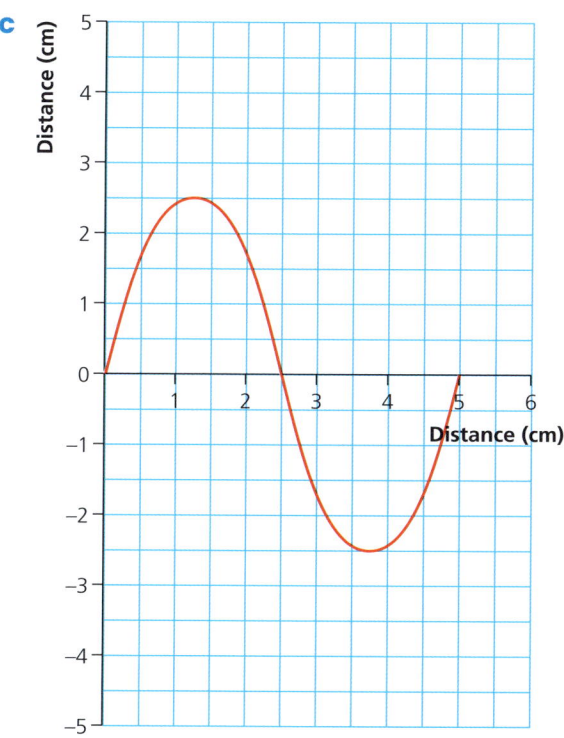
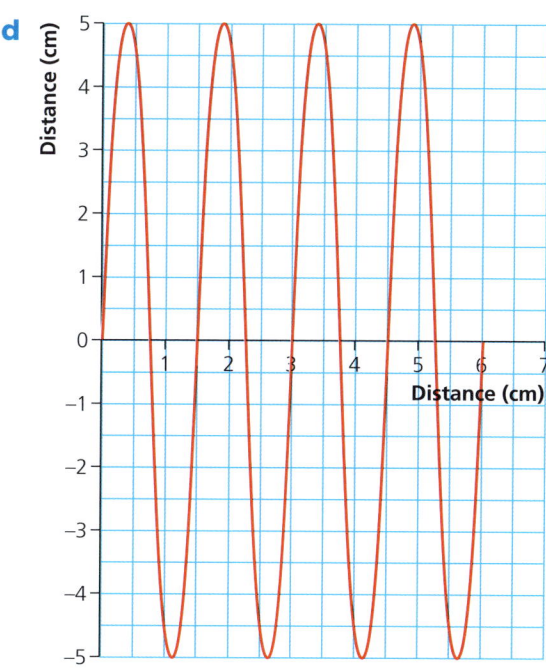

7   How do you find the amplitude of a wave?

8   How do you find the wavelength of a wave?

9   What happens to light if we increase the amplitude?

10 What is the definition of a *medium* when talking about waves?

11 You might see waves on a pond when a duck lands in it. Here, though, we are thinking about the forces on the duck as it floats on the water and swims across the pond.

    a   What is the name of the force that pulls the duck towards Earth?

    b   What is the name of the force that stops the duck sinking in water?

    c   What is the name of the force produced acting against the direction that the duck moves when the duck moves through the water?

How fast can light travel?

12 Where does light travel fastest?

13 What happens to the speed of light when it travels from a gas into a liquid?

14 Why can light travel through a vacuum when sound cannot?

15 Overall, which travels faster: sound or light?

16 Describe the trend in the table below.

Medium	Density of medium (kg/m³)	Speed of light (m/s)
Vacuum	0	299 790 000
Air	1	299 700 000
Oil	700	250 000 000
Water	1000	225 000 000
Glass	2500	200 000 000

17 How is a longitudinal wave different to a transverse wave?

18 What type of wave is sound?

19 What units do we measure frequency in?

20 What happens to a sound if we decrease the frequency of the sound wave?

21 Thunder and lightning happen at the same time. But when we are far away, we see the lightning long before we hear the thunder. Why is this?

22 Here is a line of inquiry: How does the power of a kettle affect the time it takes to boil water?

    a What is the independent variable?

    b What is the dependent variable?

> **Tip**
>
> See 'Lines of inquiry and variables' on page 7 of your Knowledge Book.

23 A person stands at the front of a room and walks all around the edge of the room until they are back where they started. Is this an oscillation? Explain your answer.

24 Sound travels much slower than light. A guitar makes sounds when its strings vibrate, and they behave elastically.

    a What is the relationship between force and extension according to Hooke's law?

    b What do we call the force required to extend or compress an elastic object by 1m?

    c In words, what equation links force, spring constant and extension? (It is also known as Hooke's law.)

    d When work is done to stretch an elastic band, what energy transfer is this?

What is light?

25 What is oscillating in a light wave?

26 What path does light take to travel from A to B?

27 Does light travel slowest in solids, liquids or gases?

28 How do we represent the path that light takes when we draw it?

29 What happens to the speed of light when it travels from a gas into a vacuum?

30 What type of wave is light?

31 A water wave is passing by. We measure the oscillations of a floating buoy (a kind of big plastic ball).

  a The buoy oscillates two times a second. What is the frequency?

  b The buoy oscillates one time a second. What is the frequency?

  c The buoy oscillates three times in 2 seconds. What is the frequency?

  d The buoy oscillates 23 times in 10 seconds. What is the frequency?

32 What do we call a region where there is no light when it is blocked by another object?

33 Copy and complete this diagram below to show the path that light takes to get from a star to your eye.

34 Copy and complete the diagram below to show the size of the shadow that is formed on the wall by the object in between the torch and the wall. Start all the rays of light from the spot.

Draw all lines starting here.

wall

35 Write as many similarities between light waves and sound waves as you can.

36 Look at the ray diagrams below. In each one, there is at least one mistake. Explain what each mistake is.

a    **b**

37 A person is looking at a phone screen. How does the phone screen transfer energy to the person's eyes?

38 Draw a ray diagram for the person looking at a phone screen.

39 How can we show forces on a force diagram?

40 In a force diagram, what does the direction of an arrow show?

41 In a force diagram, what does the length of an arrow show?

How can light waves differ?

42 If we increase the amplitude of a light wave, what happens to the light?

43 How can we change the colour of light?

44 Which colour of light has the longest wavelength?

45 Which colour of light has the shortest wavelength?

46 How does the colour of light change if the wavelength of a light wave is increased?

47 What happens to sound if the amplitude of a sound wave is decreased?

48 What do we need to do to make a light dimmer?

49 We use straight arrows to show the direction that light travels.

   a Why do we use a straight arrow to show the direction that light travels?

   b What else do we use arrows to represent in physics?

50 How can we alter a sound wave to increase the pitch?

51 What is the name for the vibrating membrane in a microphone?

52 How does a guitar produce a sound?

53 A student says, 'When I turn the brightness of my bike light up, the wavelength of the light is getting longer.' Explain why they are wrong.

54 A different bulb has a power of 20 W and is left on for 5 minutes. How much energy does it transfer?

55 What do we call the turning effect of a force?

56 When is a see-saw balanced?

> **Tip**
>
> For help, see the Worked example in Topic P1.3 of your Knowledge Book. Remember the units.

What happens to light when the medium changes?

57 What three things can happen to light when it hits a new material?

58 What is oscillating in a light wave?

59 Draw three diagrams to show what we mean when we say each of the following.

    a Light is transmitted by glass.

    b Light is absorbed by black wallpaper.

    c Light is reflected by a mirror.

60 What does *transmit* mean?

61 A student says, 'When light is absorbed, the energy disappears, so the light cannot get through to the other side.'

    a Why are they wrong?   **b** Correct their statement.

62 Why do we use an arrow to represent forces?

63 What is faster: light or sound?

64 Imagine we shine a torch through a glass fish tank filled with water. The light wave travels from air to glass to water to glass and then back to air. Describe what happens to the speed of light as it travels.

65 Light travels at nearly 300 000 000 m/s in a vacuum. What does this mean?

66 Imagine we pulse a sound through a glass fish tank filled with water. The sound wave travels from air to glass to water to glass and then back to air. Describe what happens to the speed of sound as it travels.

67 A student writes the following line of inquiry: 'Smaller people are slower.'

    a Why is this not a line of inquiry?

    b Write a proper line of inquiry to test their idea.

    c What is the independent variable?

    d What is the dependent variable?

> **Tip**
>
> See 'Lines of inquiry and variables' on page 7 of your Knowledge Book.

68 Write as many differences between light waves and sound waves as you can.

69 What is the law of conservation of energy?

70 What two things happen to light when it hits a block of wood?

71 A student shines a torch on a chair. After a little while, the temperature of the chair has gone up a tiny bit. Explain why this has occurred.

72. What do we call two forces that are equal in size and opposite in direction, while acting on a single object?
73. What is the relationship between the size of a force and the length of the arrow that represents it?
74. What is the name of the overall force when you add all the forces acting on an object?

P6.2 Interactions of light waves with materials

Why do different materials look different?

75. What happens to light when it hits an opaque material?
76. Name some materials that are opaque.
77. What happens to light when it hits a translucent material?
78. What does *emit* mean?
79. Name some materials that are translucent.
80. What happens to light when it hits a transparent material?
81. Name some materials that are transparent.
82. A student turns up the brightness on their phone screen. What is the effect on the light waves that it emits?
83. What is a vacuum?
84. Name some materials that light can travel through.
85. Measure the amplitude and wavelength of each of the waves below.

a
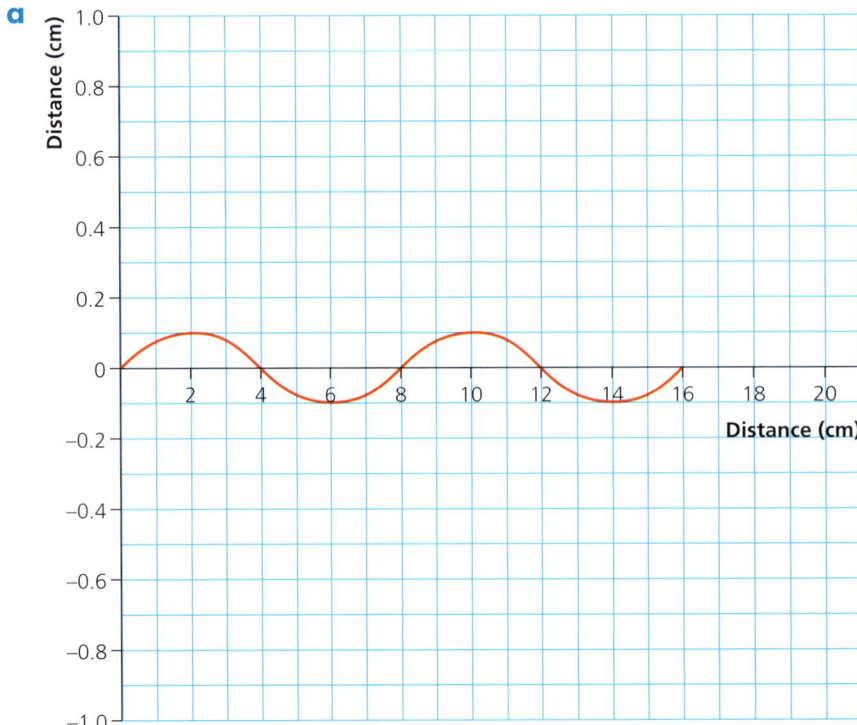

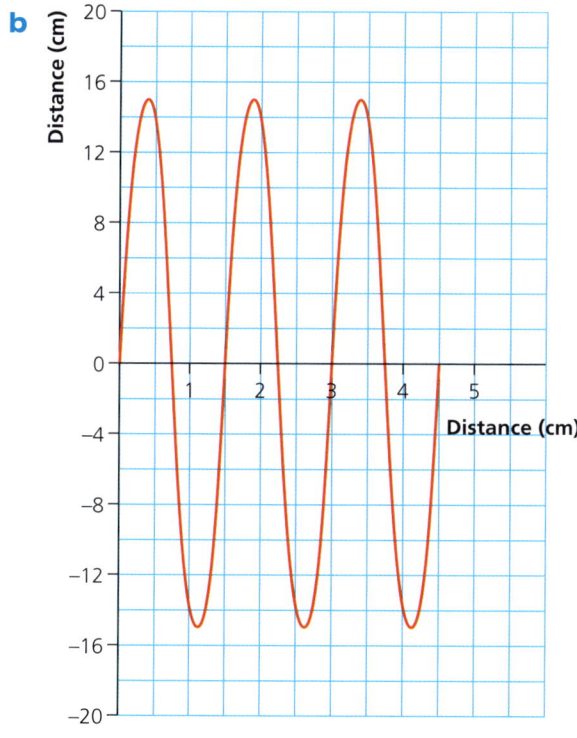

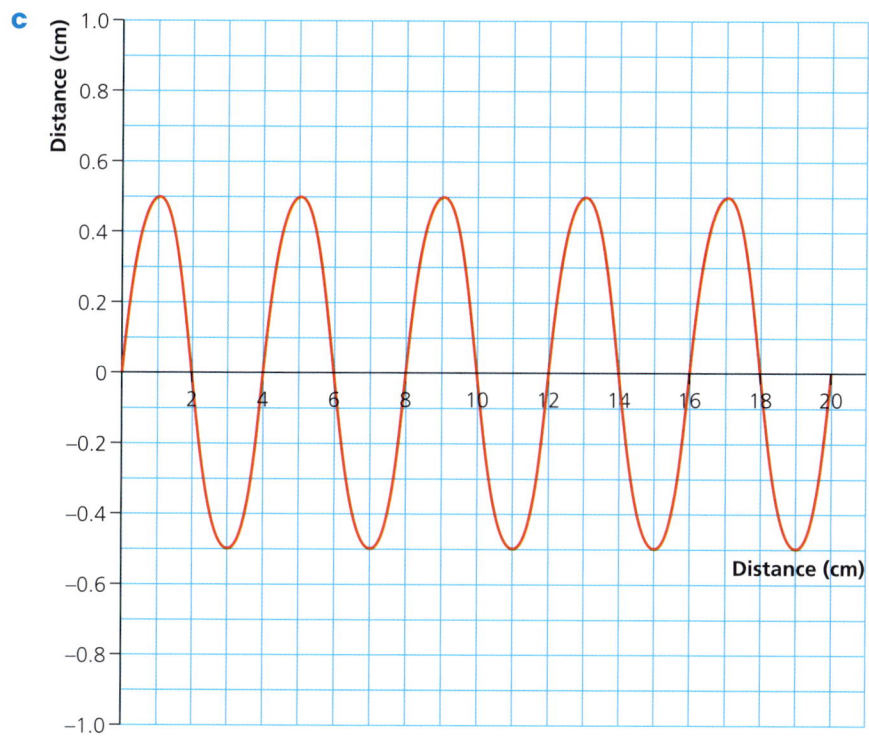

d

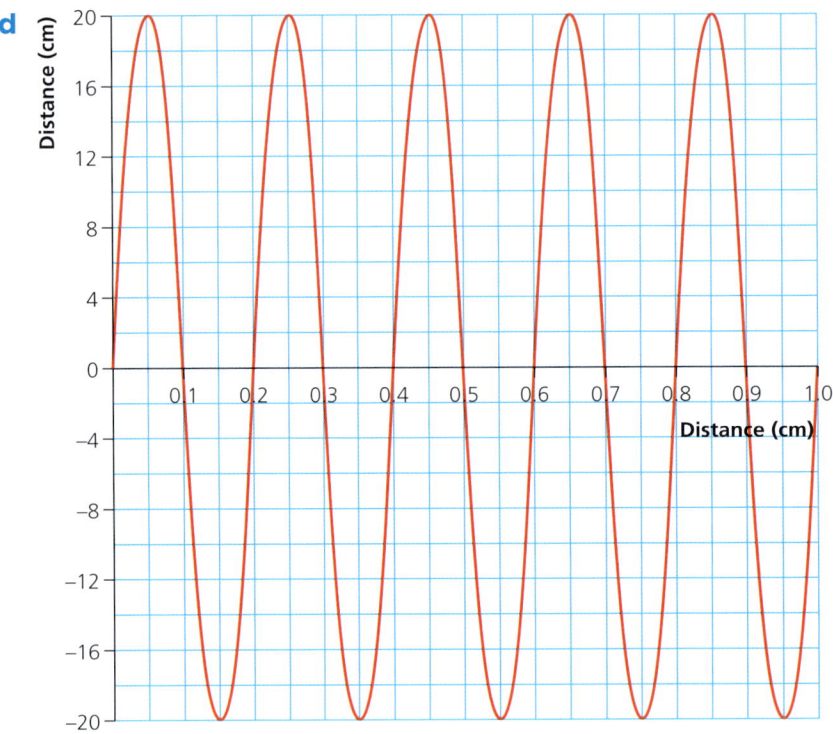

86 What is the definition of a *medium* when talking about waves?

87 We do not use the words *transparent*, *translucent* or *opaque* in relation to sound. But if we did, what would you hear if:

　a sound was played on the other side of a transparent medium

　b sound was played on the other side of a translucent medium

　c sound was played on the other side of an opaque medium?

88 A student plotted the following graph from the data in the table below. Identify the three mistakes the student has made.

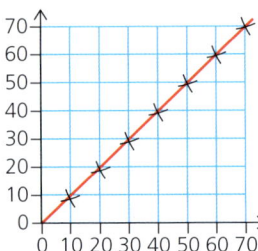

Mass (kg)	Power (W)
10	10
15	20
20	30
25	40
30	50
35	60
40	70

89 A light bulb transfers 40 J of energy towards a piece of glass, but only 35 J is transmitted through the glass. How much is not transmitted by the glass?

90 A student has a block of wood, a block of glass and a pair of sunglasses. State which one is transparent, which one is translucent and which one is opaque. Explain your answer.

What happens to light if it hits a new medium at an angle?

91. What name do we give to the process where light changes direction as it enters a new medium?
92. What happens to the speed of light when it goes from being in a gas to being in a solid?
93. What happens to the speed of light when it goes from being in a solid to being in a gas?
94. What name do we give to a material that lets none of the light through?
95. When does light refract?
96. What happens to the direction of light during refraction?
97. Why does some of the light reflect during refraction?
98. What name do we give to a material that lets some of the light through?
99. Explain whether refraction is an example of transmission, absorption or reflection.
100. What name do we give to a material that lets all the light through?
101. How many times will refraction occur when light passes from the air into a glass fish tank filled with water and out the other side?
102. A student uses a battery-powered torch to view a brick. The temperature of the brick increases. The student says, 'The temperature has gone up because the chemical energy store is now full.' Explain why the student is wrong.
103. If a window does not transmit all the light that hits it, what happens to the rest of it?

Why does refraction happen?

104. What causes light to change direction when it changes medium?
105. Why does refraction not happen when light enters a new medium at 90°?
106. What is the normal line?
107. For each of the boundaries below, say whether light will speed up or slow down, and whether it will bend towards or away from the normal.

   a Air to glass
   b Air to water
   c Water to air
   d Water to glass
   e Glass to water
   f Air to vacuum
   g Vacuum to water
   h Glass to air
   i Glass to vacuum

108 A student says, 'Light cannot travel in a vacuum as there are no particles.' Explain why they are wrong. Correct their answer in as many ways as possible.

109 Copy and complete the diagram below to show the path that light will take through the glass block.

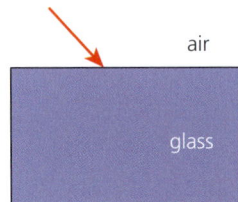

110 Explain why light bends towards the normal when it travels from water into glass.

111 Light is a transfer of energy by waves.

   a What do waves not transfer?

   b What is the law of conservation of energy?

   c If light is absorbed by an object, what happens to the energy?

112 Copy and complete the following three refraction diagrams.

a
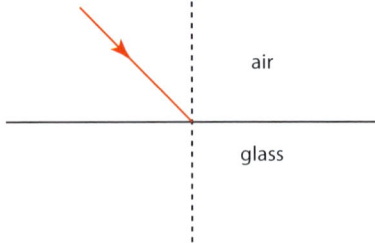

c
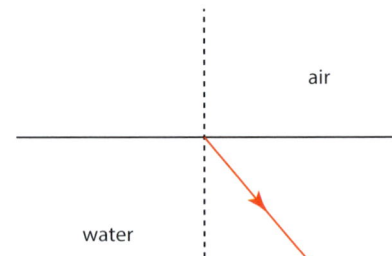

b
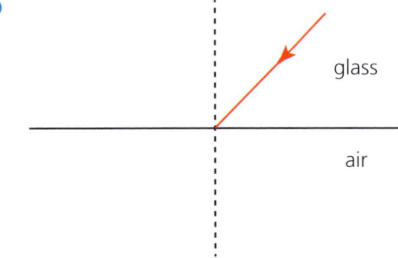

113 For each of the boundaries below, say whether light will speed up or slow down, and whether it will bend towards or away from the normal.

   a Air to plastic

   b Plastic to water

   c Vacuum to air

   d Plastic to vacuum

114 It is not just light waves that refract, sound waves can refract too. Explain why sound waves can refract when they travel into a new medium.

115 Does an opaque object transmit, reflect or absorb light?

116 Some glass transmits 98% of light. Other glass transmits 93%.
   a Which glass is better for windows? Explain your answer.
   b Which glass is better for sunglasses? Explain your answer.

117 Explain why light passing from one glass block into another identical glass block does not refract.

118 What happens to the speed of light as it travels from a vacuum into water?

119 Will light moving from water to glass slow down or speed up?

120 Will sound moving from water to glass slow down or speed up?

P6.3 Mirrors, pinhole cameras and the eye

What does a lens do?

121 What does a lens do?

122 What does *focus* mean?

123 What makes a lens able to focus light?

124 Give some examples of where we see or use lenses.

125 What happens to light in a lens?

126 Why do waves refract?

127 When does light enter a new medium and *not* refract?

128 What are the units we use to measure frequency?

129 A water wave travels 200 m in 8 s. What is its speed?

130 Sometimes you can see your reflection in a window. A student says, 'This is because windows reflect some light and transmit the rest.' Explain whether or not you agree.

131 A student says, 'I saw a book in the mirror because light came from my eyes and reflected off the mirror and onto the book.' Explain why they are wrong.

132 For each of the sentences below, state whether it is referring to absorption, reflection, transmission or refraction. Sometimes it can be more than one.
   a A student looks at themself brushing teeth in a mirror.
   b A student looks at a bird through a window.
   c A student on the ground looks at fish underwater.

> **Tip**
>
> For help, see the Worked example in Topic P2.1 of your Knowledge Book.

d A student is underwater and looking at fish.

e A student on the ground looks at fish underwater but also sees an image of themselves on the surface of the water.

f A student looks at a black chair through a window. The black chair is underneath a white light.

How does the eye work?

133 In the eye, which part focuses the light?

134 In the eye, what is the name for the opening or hole through which light passes?

135 In the eye, what is the name for the part that detects light?

136 a Copy and complete the diagram below. Add the following labels.

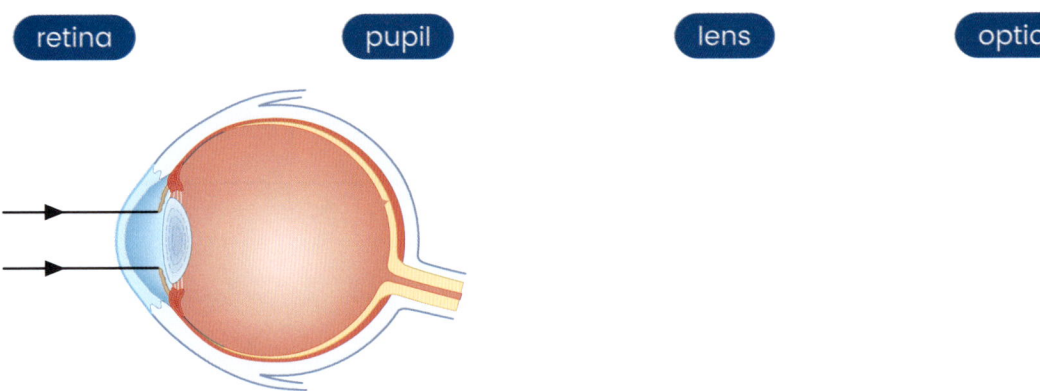

b Complete the paths of the rays of light.

137 Name the part of the eye that plays a similar role to the cochlea in the ear. Explain your answer.

138 What is the name of the process that allows a lens to focus light?

139 What is oscillating in a light wave?

140 What is oscillating in a sound wave?

141 When someone is struggling with their vision, it is because the lens in their eye is not focusing the light at the retina. Suggest why these people wear correctional lenses (glasses or contact lenses).

142 A student says, 'Lenses are translucent because they change the path of the light.' Explain why they are wrong.

143 Light travels from air into a lens, and then back into the air. Describe the changes in speed that happen.

144 What is the difference between a random and a systematic error?

145 The shape of the lens in your eye can be changed to allow you to focus on different objects.

  a How many forces are required to deform an object? Describe the directions they must be in.

  b If an object returns to its original shape and size when it is released, what do we call such an object?

146 What would happen if the optic nerve was not connected to the retina in one of your eyes?

What is a pinhole camera?

147 What is a pinhole?

148 How does light get from an object to the screen in a pinhole camera?

149 a Copy and complete the diagram below of an object and a pinhole camera. Add the following labels.

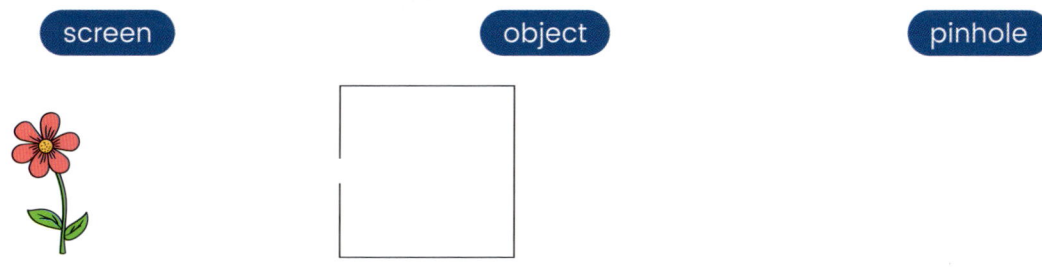

  b Complete the paths of two rays of light:
   – One from the top of the flower to the screen.
   – One from the bottom of the flower to the screen.

150 Why does a pinhole camera need no lens?

151 Why would a pinhole camera not work with a large hole?

152 Explain why the image from a pinhole camera is upside-down.

153 Why can a pinhole camera not be made of glass?

154 Copy and complete the diagram below to show the path that light will take from the football to the person's eye via reflection from the mirror. Add the following labels.

P6

155 A student draws the image below to explain how a pinhole camera works. Explain why it is wrong.

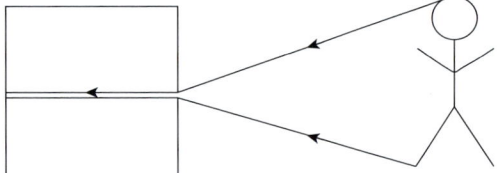

156 Measure the amplitude and wavelength of each of the waves below.

a
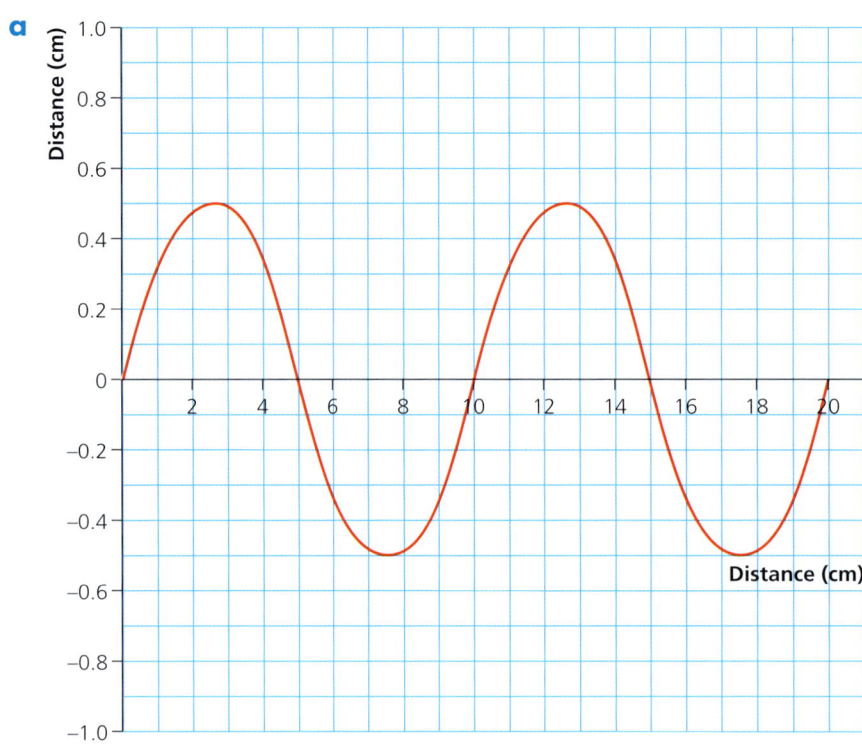

b
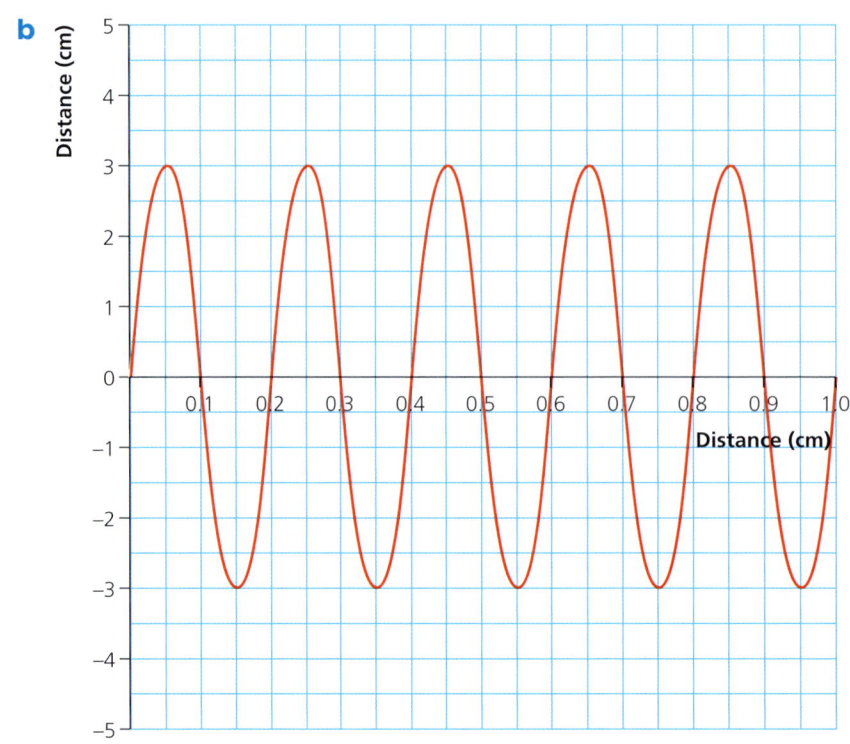

c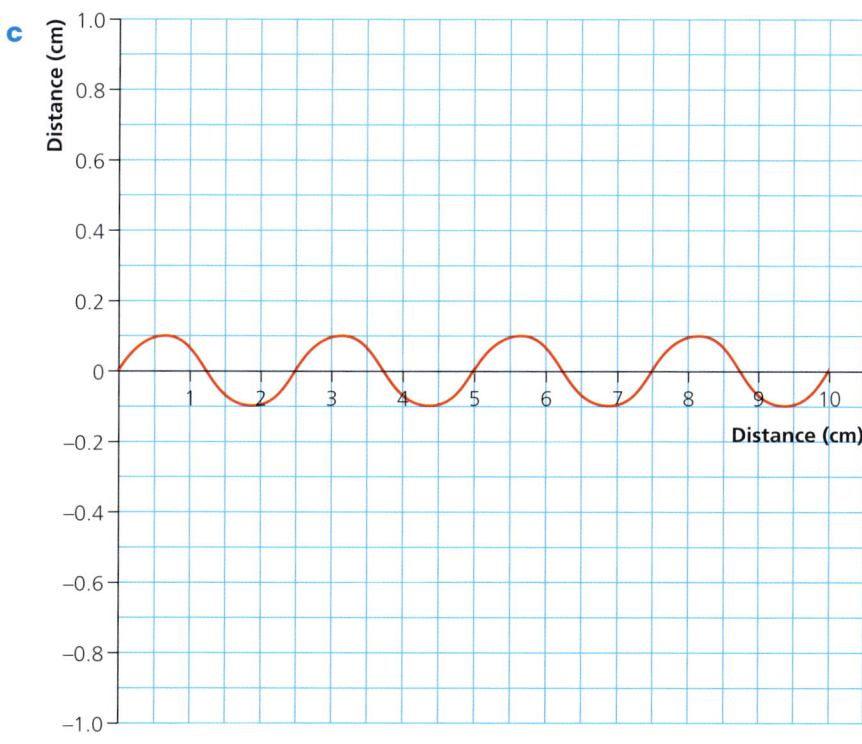

d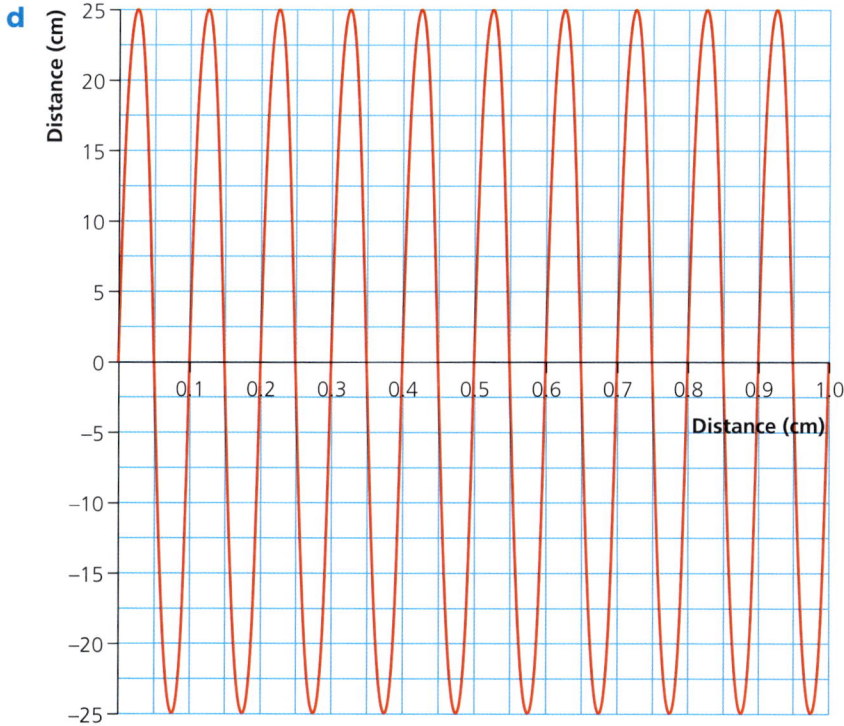

157 Copy and complete the following three refraction diagrams.

a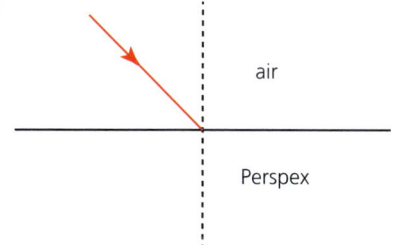

b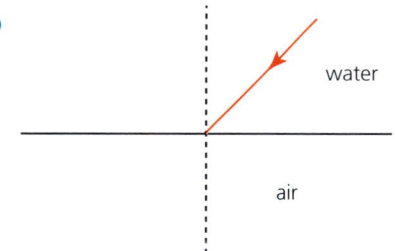

P6 Light

c

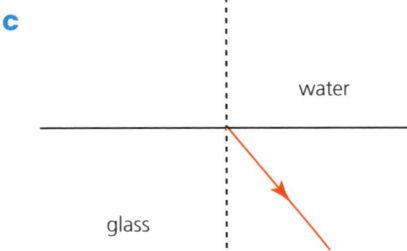

158 Draw a sketch of a pinhole camera with a screen closer to the pinhole and one with the screen further away. Explain what happens to the size of the image formed by a pinhole camera when the screen is further from the pinhole.

159 Draw a sketch of an object close to a pinhole camera and one with the object far from an identical pinhole camera. Explain what happens to the size of an image when an object moves further away from the pinhole.

160 What are the similarities between our eye and a pinhole camera?

161 What are the differences between our eye and a pinhole camera?

162 Describe as many similarities between the human eye and the human ear as you can.

163 Describe as many differences between the human eye and the human ear as you can.

P6.4 Detecting light and colour

What is a spectrum?

164 List the colours of the spectrum in order.

165 What piece of equipment can we use to split white light into a spectrum?

166 What makes one colour different from another?

167 How do we make white light?

168 Why is white light split into a spectrum of colours by a prism?

169 When white light from the Sun passes through a raindrop, the light can be split into a spectrum, creating a rainbow. In this scenario, explain what is acting as a prism.

170 What process happens to light waves to form a rainbow?

171 Which colour of light has the:

a shortest wavelength

b longest wavelength?

172 A student measures the frequency of a sound wave travelling through air.

  a The air particles oscillate 83 times a second. What is the frequency?

  b The air particles oscillate 275 times a second. What is the frequency?

  c The air particles oscillate 356 times in 2 seconds. What is the frequency?

  d The air particles oscillate 1725 times in 5 seconds. What is the frequency?

173 Name the colours A–G that should be in this diagram from top to bottom.

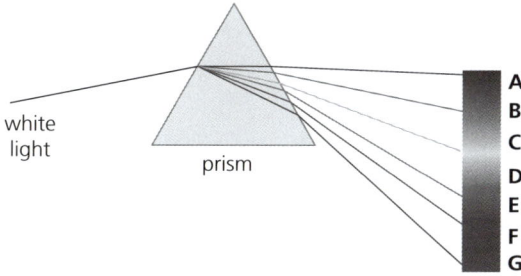

174 Name the main stores of energy.

175 We can use light from the Sun to generate electricity through solar panels. Solar panels are a renewable resource.

  a What does *renewable* mean?

  b Why can light travel from the Sun to the Earth, but sound cannot?

  c What is the transfer of energy from the Sun to the solar panel?

  d What is the transfer of energy from the solar panel to whatever it powers?

Tip

For help, see Topic P1.2 of your Knowledge Book.

176 Wind turbines use the movement of air to generate electricity by having the air turn a turbine.

  a Describe the differences between the air when it is at high pressure and when it is at low pressure.

  b Why would it be a bad idea to have wind turbines placed in areas where the air pressure is low?

P6

What makes objects the colour that they appear?

177 For each of these diagrams, say what colour the object appears.

a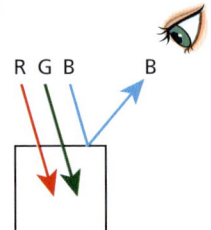

b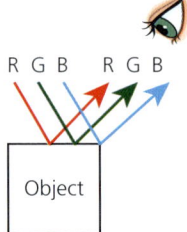

178 What colours of light (red, green or blue) does a red object:

    a reflect   **b** absorb?

179 What colours of light (red, green or blue) does a blue object:

    a reflect   **b** absorb?

180 A student says, 'A green snooker ball appears green because it absorbs green light.' Explain why they are wrong.

181 What colours of light (red, green or blue) does a black object:

    a reflect   **b** absorb?

182 List the colours of the spectrum in order.

183 What do we call a material that does not transmit any light?

184 Why do black objects tend to get warmer in the Sun than white objects?

185 A student wants to investigate how the number of motors attached to a battery affects how long the battery lasts for.

    a Write a line of inquiry for this experiment.

    b What is the independent variable?

    c What is the dependent variable?

    d What are some control variables that they need to consider?

    e Which store of energy is decreasing in the battery?

    f What is the name of the store of energy when the motor is spinning?

186 What is the resultant force when forces are balanced?

187 What three things can happen when the resultant force acting on an object is not zero?

> **Tip**
>
> For help, see the Worked example in Topic P6.4 of your Knowledge Book.

> **Tip**
>
> See 'Lines of inquiry and variables' on page 7 of your Knowledge Book.

How does the colour of light change the way objects appear?

188 For each of these diagrams, say what colour the object appears.

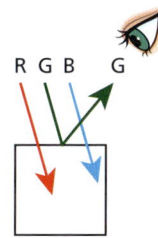

a R G B → G (eye)

b R G B → no light (eye)

> **Tip**
>
> These questions also cover 'How do filters change the way that light appears?'

189 What colours of light (red, green or blue) are let through by:

  a a red filter   **b** a green filter   **c** a blue filter?

190 What colours of light (red, green or blue) does a green object:

  a reflect   **b** absorb?

191 White light passes through a blue filter and lights up an object. What colour will the following objects appear to the human eye?

  a Red objects
  b Green objects
  c Blue objects
  d White objects
  e Black objects

192 What colours of light (red, green or blue) does a white object:

  a absorb   **b** reflect?

193 White light passes through a green filter and lights up an object. What colour will the following objects appear to the human eye?

  a Red objects
  b Green objects
  c Blue objects
  d White objects
  e Black objects

194 A student writes the following line of inquiry: 'How does the speed of a bike affect the temperature of the brakes after it stops?'

    a Draw a results table for this line of inquiry (with the heading row and one blank row for results).

    b Which store of energy increases as a bike brakes?

    c Which store of energy decreases as a bike brakes?

    d A bike has 800 J in its kinetic store and then comes to a halt. How much does the thermal store increase by?

195 A student is looking at a green tennis ball in different coloured light.

    a Explain why the tennis ball looks green to them when it is illuminated by white light.

    b How is white light made?

    c What happens to the blue portion of white light when it hits the tennis ball?

    d Explain why the tennis ball looks black to them when it is illuminated by red light.

196 What is the relationship between the size of a resultant force and the size of the change in motion?

197 When more fuel is added to a fuel tank, the chemical energy store of the fuel tank goes from 2500 J to 4500 J. How much has the chemical energy store of the fuel tank increased by?

198 A phone uses 30 J of energy from its chemical energy store when it is used for one hour. How much does it use in 20 minutes?

199 A car's petrol tank is empty. If it were to be filled with petrol, it would contain 350 000 kJ of energy.

    a How much energy is in the empty petrol tank?

    b How much energy is in the tank when it is half full?

    c Which energy store is increasing when the car is filled with fuel?

    d The car is filled up with petrol and has the engine running, but the car stays parked in the car park.

        i Is the chemical energy store of the fuel increasing or decreasing?

        ii If the chemical energy store is changing, one other energy store must be changing too. Which other store of energy changes?